Molecular and Cellular METHODS IN Developmental Toxicology

Edited by
George P. Daston

CRC Press
Boca Raton New York London Tokyo

CRC Press
METHODS IN THE LIFE SCIENCES

***Gerald D. Fasman** - Advisory Editor*
Brandeis University

Series Overview

Methods in Biochemistry
John Hershey
Department of Biological Chemistry
University of California

Cellular and Molecular Neuropharmacology
Joan M. Lakoski
Department of Pharmacology
Penn State University

Research Methods for Inbred Laboratory Mice
John P. Sundberg
The Jackson Laboratory
Bar Harbor, Maine

Methods in Neuroscience
Sidney A. Simon
Department of Neurobiology
Duke University

Joseph M. Corless
Department of Cell Biology,
Neurobiology and Ophthalmology
Duke University

Methods in Pharmacology
John H. McNeill
Professor and Dean
Faculty of Pharmaceutical Science
The University of British Columbia

Methods in Signal Transduction
Joseph Eichberg, Jr.
Department of Biochemical and Biophysical Sciences
University of Houston

Methods in Toxicology
Edward J. Massaro
Senior Research Scientist
National Health and Environmental Effects Research Laboratory
Research Triangle Park, North Carolina

CRC Press METHODS IN TOXICOLOGY

Edward J. Massaro - Advisory Editor
Senior Research Scientist
National Health and Environmental Effects Research Laboratory
Reasearch Triangle Park, North Carolina

The **CRC Press Methods in Toxicology Series** provides the reader with a step-by-step approach to each of the up-to-date methods and techniques presented in a clear and concise format. Topics covering all aspects of the methods that impact on toxicology are being reviewed for publication

Published Titles

Methods in Renal Toxicology, Rudolfs K. Zalups and Lawrence H. Lash

Molecular and Cellular Methods in Developmental Toxicology, George P. Daston

Forthcoming Titles

Methods in Inhalation Toxicology, Robert F. Phalen

PCR Application to Molecular Toxicology, Jack Vanden Heuvel

Senior Editor:	Paul Petralia
Editorial Assistant:	Cindy Carelli
Project Editor:	Sarah Fortener
Assistant Managing Editor:	Gerry Jaffe
Marketing Manager:	Susie Carlisle
Direct Marketing Manager:	Becky McEldowney
Cover design:	Dawn Boyd
PrePress:	Carlos Esser
Manufacturing:	Sheri Schwartz

Library of Congress Cataloging-in-Publication Data

Molecular and cellular methods in developmental toxicology / edited by George P. Daston.
p. cm. -- (CRC Press methods in toxicology (Boca Raton, FL)
Includes bibliographical references and index.
ISBN 0-8493-3342-3
1. Developmental toxicology--Methodology. 2. Molecular toxicology--Methodology. 3. Cytology--Methodology. I. Daston, George P. II. Series.
RA1224.45.M65 1996
618.3′2--dc20 96-14296
CIP

International Standard Book Number 0-8493-3342-3
Library of Congress Card Number 96-14296
Printed in the United States of America 1 2 3 4 5 6 7 8 9 0
Printed on acid-free paper

Abstract

This book is intended to be a laboratory manual for developmental toxicologists, at any stage of their careers, who wish to conduct research using modern molecular and cellular approaches. Experimental protocols have been selected carefully to represent the most widely used and useful techniques in molecular and cellular developmental toxicology. No other collection of protocols has been designed specifically for and by developmental toxicologists. Detailed descriptions and sources of reagents are provided. The book is intended to be taken into the laboratory and used at the bench.

Preface

The field of developmental toxicology is undergoing a revolution in its understanding of the nature of chemically induced birth defects. At the heart of the revolution is the availability of methodology that makes it possible to investigate effects at the cellular and molecular levels, the levels at which the critical events leading to toxicity occur. The purpose of this book is to describe in detail some of the more commonly applied and useful techniques for addressing mechanistic questions in developmental toxicology and teratology.

It is hoped that the reader will view this book as a practical guide. Each of the chapters describes in a clear manner the theory of — and gives detailed procedures for — many of the commonly used techniques in developmental toxicology. Sources of artifacts and appropriate controls are indicated. Many of the chapters include listings of sources of materials. Continual improvements in technology have made many of the methods presented, particularly those of molecular biology, increasingly accessible. Kits are available for most commonly used procedures. Methodological improvements have also quickened the pace of research in these areas; many procedures that once took 2 days to complete can now be accomplished in 2 hours. Where there is a choice of methods that provide comparable results, the faster and easier methods are presented.

This book was based on a continuing education course given at the annual meeting of the Teratology Society in 1994. Most of the contributors to this volume lectured in that course, although the chapters herein have been updated. The authors are among the leaders in applying molecular and cellular techniques to the study of normal and abnormal development. They know first-hand the best ways to achieve good results with embryonic specimens, and the methods they describe have been painstakingly perfected in their own laboratories. If their advice is followed carefully, success is likely.

The book is organized functionally into three sections that describe molecular, biochemical, and cellular techniques that have been applied to the study of normal and abnormal developmental processes. The first five chapters describe techniques of molecular biology, including the isolation and characterization of nucleic acids, *in situ* hybridization, transgenic animals, nucleic acid amplification techniques, and the use of antisense RNA in the study of abnormal development. The next two chapters present techniques to detect specific proteins and to localize them within the embryo. The final chapters provide methods for analyzing cell death, cell function by flow cytometry, and cell fate within the embryo.

George P. Daston

The Editor

George P. Daston, Ph.D., is Principal Research Scientist at the Miami Valley Laboratories of the Procter and Gamble Company. He earned his Ph.D. in Biology (Teratology) from the University of Miami in 1981 and has completed additional postdoctorate work in Developmental Toxicology. From 1983 to 1985, Dr. Daston was an Assistant Professor, Department of Biological Sciences, at the University of Wisconsin-Milwaukee. Since then, Dr. Daston has held various positions at Procter and Gamble, including Staff Scientist (Developmental Toxicology) and Group Leader (Developmental, Reproductive, and Neurotoxicology).

Dr. Daston's research interests include teratogenic mechanisms, *in vitro* methodologies, and risk assessment. He has published over 50 peer-reviewed articles, reviews, and book chapters. His most recent research includes (1) toxicant-nutrient (especially zinc) and maternal-embryonal interactions in developmental toxicity, (2) the role of pattern formation genes in abnormal development, (3) the use of primary cell cultures as developmental toxicant screens, and (4) improvements in risk assessment methodology for noncancer endpoints.

Dr. Daston's activities in professional societies include serving as chair of the Reproductive and Developmental Effects Subcommittee of the American Industrial Health Council; on the steering committee of the Institute for Evaluating Health Risk's Developmental and Reproductive Toxicology Project; as president (1994 to 1995) of the Society of Toxicology's Reproductive and Developmental Toxicology Specialty Section; as secretary (1994 to 1997) of the Teratology Society; and as a member of the National Academy of Sciences Board on Environmental Studies and Toxicology (BEST). Dr. Daston has recently served on the organizing committees for ILSI/EPA/AIHC workshops on benchmark dose methodology and human variability in toxic response, on an EPA workshop on endocrine-mediated toxicity, and as co-chair of an AIHC/EPA workshop on Leydig cell tumors.

Dr. Daston is an associate editor of *Teratogenesis, Carcinogenesis and Mutagenesis;* is on the editorial board of *Fundamental and Applied Toxicology* and *Human and Ecological Risk Assessment;* and is an ad hoc reviewer for *Teratology, Reproductive Toxicology, Journal of Nutrition,* and other journals. Dr. Daston is also an adjunct Associate Professor in the Department of Pediatrics and Developmental Biology Program at the University of Cincinnati and Children's Hospital Research Foundation and lectures in courses on teratology, developmental biology, toxicology, and risk assessment. Dr. Daston was a Visiting Scientist at the Salk Institute, Molecular Neurobiology Laboratory, from 1993 to 1994.

The Contributors

Urs Albrecht, Ph.D.
Research Associate
Department of Biochemistry
Baylor College of Medicine
Houston, Texas

Michael R. Blackburn, Ph.D.
Postdoctoral Fellow
Department of Biochemistry
Baylor College of Medicine
Houston, Texas

Diana K. Darnell, Ph.D.
Postdoctoral Fellow
Department of Neurobiology and Anatomy
University of Utah School of Medicine
Salt Lake City, Utah

George P. Daston, Ph.D.
Principal Research Scientist
Miami Valley Laboratories
Procter & Gamble Company, *and*
Associate Professor
Department of Pediatrics and Developmental Biology Program
University of Cincinnati and Children's Hospital Research Foundation
Cincinnati, Ohio

Gregor Eichele, Ph.D.
Professor
Department of Biochemistry
Baylor College of Medicine
Houston, Texas

Kenneth H. Elstein, M.B.A.
Research Biologist
Reproductive Toxicology Division
National Health and Environmental Effects Research Laboratory
U.S. Environmental Protection Agency
Research Triangle Park, North Carolina

Richard H. Finnell, Ph.D.
Professor
Department of Veterinary Anatomy and Public Health
College of Veterinary Medicine
Texas A&M University
College Station, Texas

Gerald B. Grunwald, Ph.D.
Professor
Department of Pathology, Anatomy, and Cell Biology *and* Department of Biochemistry and Molecular Biology
Thomas Jefferson University
Philadelphia, Pennsylvania

Jill A. Helms, D.D.S., Ph.D.
Assistant Professor
Department of Orthopaedic Surgery
University of California at San Francisco
San Francisco, California

Rodney E. Kellems, Ph.D.
Professor
Department of Biochemistry
Baylor College of Medicine
Houston, Texas

Thomas B. Knudsen, Ph.D.
Associate Professor
Department of Pathology, Anatomy, and Cell Biology
Jefferson Medical College
Philadelphia, Pennsylvania

Sally A. Little, M.S.
Research Scientist
Birth Defects Research Laboratory
Division of Congenital Defects
Department of Pediatrics and Biological Structure
University of Washington
Seattle, Washington

Hui-Chen Lu, B.S.
Research Assistant
Department of Biochemistry
Baylor College of Medicine
Houston, Texas

Scott A. Mackler, M.D., Ph.D.
Assistant Professor
Department of Medicine
Philadelphia Veteran's Affairs Medical Center
University of Pennsylvania School of Medicine
Philadelphia, Pennsylvania

Philip E. Mirkes, Ph.D.
Research Professor
Birth Defects Research Laboratory
Division of Congenital Defects
Department of Pediatrics and Biological Structure
University of Washington
Seattle, Washington

Linda Foerst Potts, Ph.D.
Department of Cell Biology and Anatomy
University of North Carolina at Chapel Hill
Chapel Hill, North Carolina

T. W. Sadler, Ph.D.
Professor
University of North Carolina Birth Defects Center
Department of Cell Biology and Anatomy
University of North Carolina at Chapel Hill
Chapel Hill, North Carolina

Gary C. Schoenwolf, Ph.D.
Professor
Department of Neurobiology and Anatomy
University of Utah School of Medicine
Salt Lake City, Utah

Scott J. Vacha, B.S.
Department of Veterinary Anatomy and Public Health
College of Veterinary Medicine
Texas A&M University
College Station, Texas

Robert M. Zucker, Ph.D.
Research Biologist
Reproductive Toxicology Division
National Health Effects and Environmental Research Laboratory
U.S. Environmental Protection Agency
Research Park Triangle, North Carolina

Table of Contents

Part III. Cellular Techniques

Dedication

To Paul and Marie Daston, my parents,
who gave me my first chemistry set.

Chapter 1

Molecular Techniques for Developmental Toxicology: An Introduction

George P. Daston

Contents

0-8493-3342-3/97/$0.00+$.50

The technology of molecular genetics has revolutionized research in a large number of fields, including developmental toxicology. It is now possible to directly address questions about the effect of xenobiotic chemicals on the expression and function of the genes controlling development. This ability to conduct research at the most basic level of biological organization is already dramatically changing our understanding of the control mechanisms of development and how they are perturbed, leading to congenital malformations.

The armament of molecular techniques is truly vast and beyond the scope of this chapter or even this book. There are comprehensive methods anthologies available, and these are listed at the end of this chapter. In recent years, kits have been assembled that provide the investigator with a complete set of reagents and written procedures for accomplishing a specific task. I have not provided extensive protocols for procedures that can be better done by using a kit. A list of suppliers of molecular biology equipment, supplies, and kits is given in Appendix B.

I. Molecular Biology: The Basics

Deoxyribonucleic acid (DNA) is the genetic material of cells. It encodes all of the information specifying the primary structure of all proteins in the organism. DNA is a double-stranded polymer. Each strand consists of four nucleotides — adenine (A), guanine (G), cytosine (C), and thymine (T) — linked by a backbone of covalently linked deoxyribose sugars and phosphoesters. The two strands are held together by hydrogen bonding between the nucleotides of opposite strands. This bonding is specific and complementary: A binds T, C binds G. The strands are twisted around each other (the famous double helix) and are in opposite polarity with respect to their phosphate-sugar backbones.

DNA serves as a template for making ribonucleic acid (RNA), which in turn serves as the template for the synthesis of proteins. RNA is similar to DNA except that it is single stranded, has ribose instead of deoxyribose, and has uridine (U) instead of thymine. We will be most interested in evaluating mRNA (m for messenger), because its expression dictates the kinds of proteins that a cell synthesizes and is therefore cell type and developmental stage specific. However, there are other types of RNA, particularly ribosomal (r) and transfer (t) RNA, that compose the majority of the RNA in a cell. Even though the RNA is single stranded, it will complex with a complementary strand of DNA or RNA. This makes it possible to use sequences of DNA or RNA to detect specific mRNA.

The genes in prokaryotic organisms are arranged in a continuous linear fashion such that the DNA sequence corresponds directly to the mRNA sequence. This simple arrangement is the key to many molecular techniques. Eukaryotic genes on the other hand are interspersed with regions called introns that are not translated into proteins. The regions that contain the coding sequence for proteins are called exons. In eukaryotes, a RNA transcript of both exons and introns is first synthesized, and then the introns are removed and the exons joined together to form mRNA. In a final processing step, almost all mRNA has a poly-A tail added to the 3′ end before it leaves the nucleus. This unique feature can be utilized to separate mRNA from total RNA when that is desirable, by techniques that will be described later in the chapter.

One of the most important early discoveries of molecular biology was that, in addition to DNA serving as a template for RNA, certain viruses were able to make DNA from a RNA template using the enzyme reverse transcriptase. The DNA made in this way is called cDNA (c for complementary). Bacterial restriction enzymes are another important discovery in molecular genetics. These enzymes cut DNA at specific nucleotide sequences. Hundreds of these enzymes have been discovered, and a large number of target sites are available. It is possible to pick and choose among these to cut out a specific region of DNA whose sequence is known. It is also possible to cut the entire genome into manageable pieces for the construction of a library of DNA.

Libraries of DNA can be constructed from the entire genome of an organism by treating cellular DNA with restriction enzymes (genomic library). Or, a library of cDNA made by reverse transcriptase can be made only from the mRNA being expressed by a particular cell type or tissue. In either case, the DNA is inserted into vectors, which are DNA molecules that can be introduced into a host cell (usually a bacterium) where the vector along with its insert can be propagated. This procedure is known as cloning, and the host cell is called a clone. The most commonly used vectors are plasmids, circular pieces of DNA that are replicated extrachromosomally in bacteria. Plasmids have been engineered so that they have multiple features that make it easy to insert and transcribe cDNA and to select clones that contain a plasmid. A typical plasmid vector, marketed by Stratagene, is shown in Figure 1.1. It contains:

1. A multiple cloning site, with a large number of different restriction enzyme recognition sites to facilitate the introduction or removal of the cDNA
2. An antibiotic resistance gene, so that only the bacteria that have been transformed by introduction of a plasmid can survive antibiotic treatment
3. A *lac Z* (β-galactosidase) gene, to allow for blue-white selection: the gene is interrupted by introduction of DNA into the multiple cloning site, and the β-galactosidase protein produced is nonfunctional; the clones can be incubated in the presence of a colorless substrate that is made into a blue product by functional β-galactosidase, allowing one to determine whether the plasmid has a DNA insert; many vectors including this one also contain a *lac I* repressor gene that can be chemically inhibited to enhance the contrast between blue and white colonies

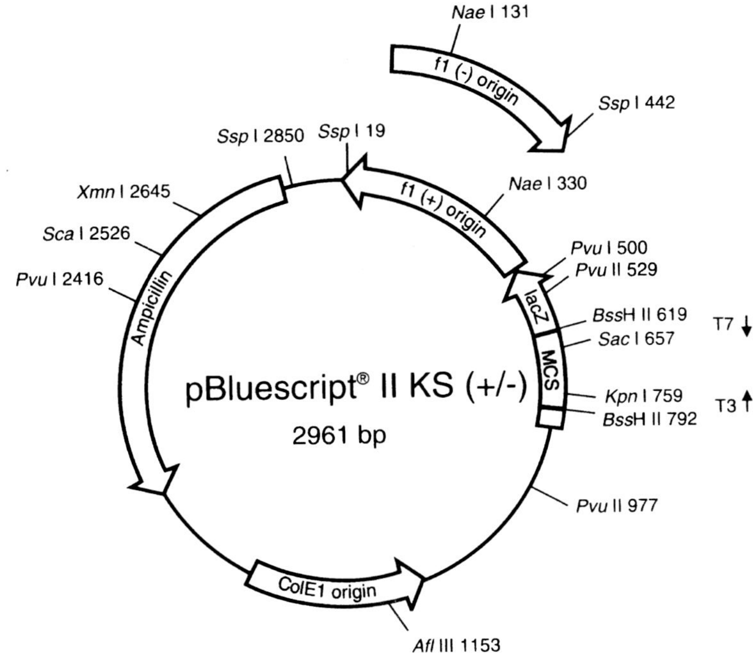

FIGURE 1.1

A typical plasmid vector. It has a multiple cloning site (MCS), an ampicillin resistance gene, a *lac Z* gene for blue-white selection, and promoters flanking the insertion site (T3 and T7 for T3 and T7 RNA polymerases) so that both strands of cDNA can be transcribed. This plasmid is sold by Stratagene. (Figure courtesy of Stratagene Technical Services.)

4. Promoter regions flanking the insert site: this feature allows RNA polymerase to transcribe either strand of the cDNA (the two promoters bind different polymerases) and allows the synthesis of RNA that is either complementary (anti-sense) or identical (sense) to the original mRNA from which the clone was derived

In addition to these features, plasmids are easy to isolate from the host bacteria and are easy to sequence. Disadvantages are that it is still tedious to screen plasmid libraries and the length of the insert must be relatively short, less than 10 kilobases (Kb) of DNA. Other vectors that are often used include bacteriophage lambda, cosmids (hybrids of phage and plasmid), and yeast artificial chromosomes, which can accept inserts of 1000 Kb or more. For the purposes of generating large quantities of RNA probes from cDNA that has already been isolated, however, plasmids are ideal.

Most novices will not be involved in cloning and sequencing libraries, so protocols for these will not be included here. Perhaps the most widely applicable techniques for developmental toxicology are those for measuring the expression of specific mRNAs. These techniques use cRNA probes that are transcribed from plasmids. The procedures for making the probes vary somewhat depending on the identity of the promoter region, so there is no generic protocol provided. Most vendors of plasmids also sell kits for the synthesis of probes from that plasmid. The kits come with specific instructions and the appropriate enzymes and reagents. Sometimes cDNAs are in plasmids that do not permit cRNA synthesis, so a procedure for inserting cDNA into a vector appropriate for making RNA probes is included. The rest of this chapter will be devoted to techniques for the isolation of RNA from tissue and quantitation of specific mRNAs.

II. Some General Rules

The single most important point to remember is that good results can be obtained only if one is fastidious, particularly when working with RNA. Ribonucleases (RNases), enzymes that degrade RNA, are ubiquitous and can quickly ruin a sample. It is imperative that gloves always be worn when working with RNA, as skin is a good source of RNases. Disposable plastic labware should be used whenever possible, as glass tends to harbor RNases. RNases are very difficult to denature, but glassware can be used after it has been baked at 150°C for 4 hours. Distilled water used as is or to make solutions should be treated with diethylpyrocarbonate (DEPC), an inhibitor of RNase, and then autoclaved. The final concentration of DEPC should be 0.1%. The notable exception is solutions buffered with Tris, which should not be treated with DEPC because the two react. In addition to these precautions, it is desirable to set aside reagents for procedures involving RNA so that they are not used for other procedures. These reagents should be stored separately and only handled by personnel wearing gloves. Spatulas used for scooping out solid reagents should be soaked in DEPC-treated water and autoclaved.

It is standard practice to label nucleic acid probes with a radioisotope for the purpose of detecting the probe. The most commonly used radiolabel for the techniques described in this chapter is ^{32}P, which is a strong emitter of radiation, but the β-particles that are emitted can be stopped effectively with Plexiglas shielding. One should only work with ^{32}P using appropriate shielding. A well-equipped laboratory also should have a Plexiglas container for storing unused isotope and a large Plexiglas box for temporarily storing ^{32}P-contaminated waste. ^{32}P has a half-life of less than 2 weeks, so it should not be ordered until one is ready to synthesize probes. There are products on the market that substitute chemiluminescent labels for radioactivity (e.g., Boehringer Mannheim's GENIUS® products). These are probably suitable for the applications described in this chapter and should be seriously considered by those whose labs are not equipped for radioisotope research.

There is a set of basic equipment that every lab engaged in molecular biology needs. The most important factor to consider is that most reactions and procedures are carried out in extremely small volumes because of the expense of many of the reagents and the very small quantities of sample available. Therefore, it is necessary to have equipment capable of handling small volumes: automatic micropipettors (e.g., Pipetman brand) capable of accurately dispensing volumes of 1 μl or less and table-top microcentrifuges for 1.5-ml microcentrifuge tubes. It is preferable to have at least two microcentrifuges, one for room temperature applications and another that is refrigerated (or a nonrefrigerated model can be kept in a cold room or chromatography cabinet). At least two small water baths that can be set at different temperatures are also required. An apparatus for horizontal gel electrophoresis and a power supply are also required for separating nucleic acids on agarose gels. Buy smaller units, and buy more than one Plexiglas gel support so that while one gel is running another can be cast. A microwave oven for melting agarose is also useful. A vertical gel apparatus for polyacrylamide gels is needed for the RNase protection assay described later.

In addition to the equipment listed above it will be necessary to have access to equipment for growing bacteria, including a shaking incubator, a static incubator, and an autoclave. One will also need access to an ultraviolet (UV) spectrophotometer, a dark room, a UV light box, a scintillation counter, centrifuges, and other equipment that generally is shared among laboratories.

III. Preparing Probes

Probes for detecting specific mRNAs can be made of DNA or RNA. cRNAs can be made much more radioactive (and therefore more sensitive), and it is generally easier to make them. In order to make cRNA probes, the cDNA must be in a vector that contains promoter regions for DNA-dependent RNA polymerases, such as the T3 and T7 RNA polymerase sites in the vector pictured in Figure 1.1. If you are fortunate, the cDNA that you have obtained is already in one of these vectors. If not, the cDNA can be removed from its vector and inserted into a more appropriate vector, a procedure known as subcloning. Subcloning is simple if the restriction enzyme recognition sites present on the cDNA are also present on the plasmid. (You should ask the person providing you with the cDNA for its sequence or a restriction enzyme site map.) Then all that needs to be done is for the insert and plasmid to be cut with restriction enzymes and then for the ends of the plasmid and vector to be ligated enzymatically. If the new vector has no corresponding restriction enzyme sites to the cDNA, then both DNAs need to be prepared further by filling in the recessed ends of the restriction-enzyme-cut DNA before ligation. The procedure is as follows:

A. Restriction Enzyme Cut of DNA

1. Place 5 μl of a 1-mg/ml solution of DNA into a 1.5-ml microcentrifuge tube.
2. Add: 5 μl 10× restriction enzyme buffer (supplied with the enzyme)
   5 μl restriction enzyme
   35 μl water
3. Incubate at 37°C for 1 hour.

Select restriction enzymes that do not also have sites within the insert or elsewhere on the vector. Note that the restriction enzyme should never constitute more than 10% of the mix. If it is necessary to use two restriction enzymes to cut both ends of the cDNA, then use only 2.5 µl of each. The buffer provided with the restriction enzyme is optimized for that enzyme and not all restriction enzymes work in all buffers; therefore, consult the chart of buffers provided by the restriction enzyme supplier to select a buffer that will be compatible with both enzymes. This may mean that neither enzyme functions optimally, and the incubation time may need to be increased. Alternatively, it is possible to add each enzyme sequentially, with its own optimum buffer. Enzymes requiring high-salt buffers always should be added last.

B. Isolation of cDNA Fragment

1. Extract the DNA from the restriction enzyme mix:
   a. Add 1 volume of salt-saturated phenol (Appendix A) and 1 volume of chloroform
   b. Vortex
   c. Centrifuge for 1 minute
   d. Remove the upper phase to a new tube, discarding the lower phase and interphase
   e. Precipitate the DNA by adding 0.1 volumes of 3 *M* Na acetate (pH 5.2) and 2.5 volumes of 95% ethanol; cool at –20°C for 15 minutes; then centrifuge for 15 minutes at 4°C
   f. Dissolve the precipitate in 5 µl TE buffer (Appendix A)
2. The cDNA insert then can be isolated from the plasmid by electrophoresis.
   a. Make a 1% solution of low melting point agarose in TAE buffer (Appendix A). (Do not use TBE buffer, as the borates will inhibit the ligases to be used later.) Low-melting-point agarose is different from regular agarose, which will not work well in this procedure.
   b. Microwave the agarose just to boiling, mix, then add 1.5 µl of a 10-mg/ml solution of ethidium bromide.
   c. Pour molten agarose into a horizontal gel electrophoresis apparatus. The amount of agarose solution needed depends on the size of the apparatus used but is usually less than 50 ml.
   d. *qs* the DNA solution to 15 µl with TE buffer.
   e. Add 3 µl of a running buffer consisting of 0.25% bromphenol blue in 30% glycerol.
   f. Load the sample into a well in the agarose gel after it has solidified. Also load molecular weight markers (available commercially) of the same size range as the cDNA insert.
   g. Connect an electrophoresis power supply and run at 5 V/cm of gel width (usually about 40 V). Nucleic acids run toward the positive (red) terminal of the electric field. (Remember: "run to red.") Run until the bromphenol blue tracking dye is within 1 to 2 cm of the end of the gel.

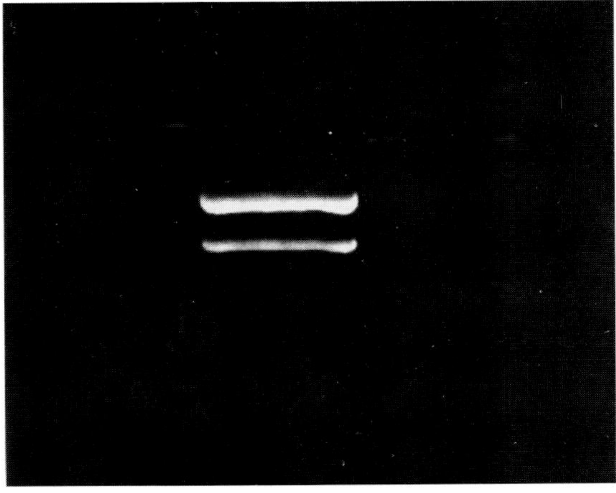

FIGURE 1.2
An agarose gel stained with ethidium bromide and illuminated with ultraviolet light. The two bands are the result of cutting a plasmid with restriction enzymes. The upper band is the higher molecular weight vector DNA and the lower band is the excised cDNA insert.

h. Visualize the DNA by placing the gel on an ultraviolet light box (use appropriate eye protection). Two bands should be visible, as in Figure 1.2. The larger band, closer to the loading well, is the plasmid; the smaller is the insert. Check that the molecular weight of the insert corresponds to the appropriate molecular weight standard. If there are more than two bands, it is likely that there were other restriction enzyme sites on the insert or vector in addition to the one targeted. If there are relatively few extra bands and they are all much larger than the insert, then one can be reasonably assured that these are from the vector and that the insert is intact. If there are numerous small pieces, however, it will be necessary to start again, choosing different restriction enzymes. Consult the restriction enzyme maps of both vector and insert to ensure that you have chosen unique sites.

C. Elution from Low Melting Point Agarose

1. Cut the insert from the gel using a clean razor blade.
2. Place band in a 1.5-μl microcentrifuge tube.
3. Add 30 μl 3 *M* Na acetate (pH 5.2).
4. Add 100 μl TE buffer.
5. Heat at 70°C for 10 to 15 minutes.
6. Immediately add 200 μl salt-saturated phenol.
7. Vortex continuously for 1 minute. Repeat.
8. Centrifuge at top speed for 5 minutes.
9. Remove the supernatant to a clean tube. Repeat steps 6 through 9.
10. Add 1 volume of chloroform; vortex, and save the supernatant in a clean tube.
11. Precipitate DNA with 0.1 volume of 3 *M* Na acetate and 2.5 volumes of 95% ethanol. Cool at –20°C for at least 15 minutes, then centrifuge at 4°C for 15 minutes.

12. Reconstitute the pellet in 20 µl of TE buffer. Check the DNA content using the UV method described in the RNA extraction method below (step 7 under Section IV, RNA Extraction).

D. Filling in Cut Ends

(Do this procedure only if there are no corresponding restriction enzyme sites on the new vector.)

1. *qs* the cDNA solution to 50 µl with water.
2. Add: 5 µl 10× Klenow buffer (supplied with enzyme)
   2 µl of a 10 m*M* solution of all four dNTPs (dATP, dGTP, dCTP, dTTP)
   2 U Klenow fragment of DNA polymerase
3. Incubate at room temperature for 30 minutes.

E. Ligating the Insert and Vector

1. The vector should be prepared using the same restriction enzymes as were used to cut the insert. If there were no corresponding sites, then the vector will have to be cut at the appropriate site and treated with Klenow fragment, as in Part 4 above.
2. Place approximately 100 ng of vector DNA and 200 ng insert DNA in a 1.5-ml microcentrifuge tube; *qs* to 12.5 ml with water.
3. Add: 1.5 µl 10× ligation buffer (contains ATP)
   1 µl of T4 DNA ligase
4. Incubate overnight at 16°C.

After this step, large quantities of the new plasmid must be made by transforming bacteria (a procedure in which the plasmid is incorporated into the cytoplasm of the bacteria). There are many protocols for transforming bacteria, but here is an easy one.

1. Thaw competent (capable of being transformed) *Escherichia coli* (can be commercially purchased) on ice. Use 200 µl per transformation.
2. Add 1 to 1000 ng DNA (100 ng is optimal).
3. Vortex; put on ice for 45 to 60 minutes.
4. Incubate at 42°C for exactly 90 seconds.
5. Add 800 µl SOC medium. (SOC: 20 g bacto-tryptone, 5 g bacto-yeast extract, 0.5 g NaCl, 186 mg KCl, adjusted to pH 7.0 and autoclaved. After autoclaving, add 20 ml of a 1 *M* glucose solution that has been filter sterilized.)
6. Incubate at 37°C for 30 minutes.
7. Plate 100 µl onto LB agar plates (Appendix A) containing 60 mg/ml ampicillin.
8. Grow overnight at 37°C.

The ampicillin kills off all the bacteria except those that were made ampicillin resistant by the incorporation of the plasmid; however, not all plasmids will contain the insert, as it is possible for the cut ends of the plasmid to be ligated without the insert. Therefore, it is necessary to identify the plasmids containing the insert. This can be accomplished easily if a vector was chosen that allows for blue-white selection; i.e., the insertion of the cDNA into the plasmid interrupts the β-galactosidase gene such that the enzyme is no longer functional and cannot convert a colorless substrate into a blue product. Blue-white selection is done by adding X-gal, a substrate for β-galactosidase, and isopropylthio-β-D-galactoside (IPTG), an inhibitor of the β-galactosidase repressor gene, to the agar plates on which the transformed bacteria are grown. Add 40 μl of a stock solution of 20 mg/ml X-gal in dimethylformamide and 4 μl of a 200-mg/ml stock of IPTG, in water, to pre-poured LB agar plates before plating bacteria. (The stock solutions should be stored at –20°C, in the dark.) Colonies that have the insert will be white; others will be blue. Select white colonies for expansion of the plasmid.

Once a colony with the plasmid and insert is identified, the plasmid can be amplified by growing large amounts of bacteria by the following procedure. It is usually best to grow up several white colonies through the first step below.

1. Transfer the colony from the agar plate to a tube containing 10 ml sterile LB broth (Appendix A) containing 60 μg/ml ampicillin. Incubate overnight at 37°C with vigorous shaking. At this point, it is prudent to double-check that the bacteria being grown really have the insert and not some stray plasmid. The easiest way to check is to prepare plasmid DNA from some of the bacteria in each tube, cut the DNA with the restriction enzymes that will separate the insert from the plasmid, and run an agarose gel to ensure that an insert of the appropriate size is present. A number of protocols is available for lysing bacteria and preparing plasmid DNA, but the easiest way is to use a kit such as the Qiagen Mini-Plasmid Purification Kit. The restriction enzyme digest is done as described above, and the agarose gel procedure is comparable except that regular agarose is used. The concentration of agarose used depends on the size of the insert, but 1 to 1.5% agarose usually is appropriate. The smaller the insert, the more concentrated the gel should be. Alternatively, it is possible to evaluate the plasmid using a probe specific for the insert DNA. This more definitively identifies the insert DNA; however, the procedure described above is almost always adequate as it is extremely unlikely that a contaminating plasmid would have the same size restriction enzyme fragments as those of the correct insert and vector. Once you have assured yourself that the bacteria contain the plasmid, proceed with the next steps using only one of the overnight cultures.
2. Inoculate 25 ml of LB + ampicillin, in a 100-ml flask, with 0.1 ml of the overnight culture. Incubate at 37°C with vigorous shaking until the culture has an optical density at 600 nm of 0.6. It should take several hours for the culture to reach this point.
3. Pour the entire contents of the 100-ml flask into a 2-l flask containing 500 ml of LB + ampicillin. Incubate at 37°C with vigorous shaking until the optical density at 600 = 0.4 (approximately 2.5 hours).

4. Add 2.5 ml of a 340 mg/ml solution of chloramphenicol in ethanol.
5. Incubate 12 to 16 hours at 37°C, with vigorous shaking.
6. Isolate the plasmid DNA using a kit for this purpose, such as the Qiagen Maxi-Plasmid Purification kit.

F. Making Probes

The companies that sell the vectors suitable for cRNA probes also provide kits for synthesizing the probes. These are specific for the RNA polymerase promoter regions present on the plasmid.

IV. RNA Extraction

RNA can be extracted from whole embryos or specific structures or from maternal tissues. RNA extraction has become much simpler in recent years due to the commercial availability of a number of reagents or kits designed for this purpose. We have had good results with the RNAzol reagent from Tel-Test, Inc., and TRIzol reagent from Life Technologies. Our procedure for RNA isolation using RNAzol is as follows:

1. Isolate the embryos as quickly as possible. Homogenize the embryos in a solution of RNAzol. The instructions call for 2 ml for every 100 mg of tissue. We typically pool all embryos from a litter and assume that a rat embryo weighs 5 mg on gestation day 10, 10 mg on gestation day 11, and 25 mg on gestation day 12. For mouse embryos, we assume 5 mg on day 8, 10 mg on day 9, and 25 mg on day 10. These values are not precise, but they provide a reasonable guide for adding the reagent. If very small samples are used, there are special procedures detailed in the instructions for each kit. The embryos can be homogenized in a Polytron or equivalent, a Teflon-glass homogenizer, or any other device that will disrupt cells. For very small samples, we have used a sonicator with a microprobe with good result. If the samples cannot be processed immediately, they should be frozen in liquid nitrogen and stored at –70°C.
2. Add 0.1 ml chloroform for each ml of homogenate, cover and shake vigorously for 15 seconds, then cool on ice for 5 minutes.
3. Centrifuge at 12,000× g for 15 minutes at 4°C. The resulting solution has two phases. The upper aqueous phase has the RNA and should be saved. Do not collect the interphase, as it contains DNA and proteins.
4. Precipitate the RNA by adding an equal volume of isopropanol and cooling the samples to –20°C for at least 15 minutes. Centrifuge at 12,000× g for 15 minutes at 4°C. The RNA is a white precipitate at the bottom of the tube. It may not always be visible, depending on the size of the original sample.
5. Decant the liquid and wash the pellet in 75% ethanol by vortexing and recentrifuging as above. The amount of ethanol is not critical: 1 ml is plenty.

A. Transfer to Nylon Membrane

1. Remove formaldehyde from gel by soaking in 0.1× TAE for 1 hour.
2. Cut 10 sheets of blotter paper and one piece of nylon membrane to the size of the gel. Do not cut the paper or membrane larger than the gel. If the paper from each side of the gel touches during the transfer, current will travel through the paper, not through the gel.
3. Saturate the blotter paper with 0.1× TAE. Soak the nylon membrane in ddH_2O for 5 to 10 minutes, then in 0.1× TAE for 10 to 20 minutes.
4. Cut a Mylar mask slightly smaller than the size of the gel, as for the first layer.
5. Stack the gel sandwich on the anode as follows:
   a. Place the Mylar sheet on the anode (bottom) plate of the unit.
   b. Stack five sheets of wet blotter paper over the hole in the Mylar sheet.
   c. Place the presoaked nylon membrane on top of the blotter paper.
   d. Place the gel on top of the membrane.
   e. Place five sheets of wet blotter paper over the gel.
   f. Put the lid containing the cathode plate over the stack.

Note: *Between each layer, roll a pipette over the stack to remove any trapped air bubbles. This is especially important on the layer between the membrane and the gel.*

6. Plug the safety interlock lead from the lid into the base of the unit and connect the electrodes to the power supply.
7. Transfer at a constant current of 50 mA for 30 minutes. The voltage may increase slightly during the run.
8. Remove the upper layer of blotter paper. Make pinholes through the gel and membrane at the location of the molecular weight markers closest to that of the mRNA of interest.
9. Fix the RNA to the membrane. This can be done most easily by using an ultraviolet cross-linker, a device about the size of a microwave oven that is programmed to emit a certain amount of UV radiation to crosslink the RNA to the membrane. If this device is not available, the RNA can be attached to the membrane by drying the membrane at 80°C for 2 hours, under vacuum.

The membrane is now ready for hybridization to a cRNA probe. The hybridization is done in sealed cellophane bags. These are the same kind of bags used to seal and freeze leftovers, and both bags and sealers can be purchased from scientific supply companies or from Sears. It is important in sealing the bags that no air bubbles are trapped, as these prevent the cRNA in solution from interacting with the membrane. Air bubbles can be removed by guiding air bubbles out the unsealed top of the bag using a glass pipette or the lip of a countertop. This should be practiced several times with a dummy membrane and hybridization solution containing no radioactivity. Always seal the bag twice to prevent leakage.

B. Hybridization of RNA

1.

Hybridization Solution Components	Final Concentration	Amount
Formamide[a]	50%	15 ml
20× SSPE	5×	7.5 ml
20% Sodium dodecyl sulfate (SDS)	0.5%	750 μl
100× Denhart's solution	2×	1600 μl
Salmon Testis DNA (10 mg/ml)[b]	100 μl per 10 ml hyb. soln.	300 μl
^{32}P-cRNA probe[b]	1×10^6 cpm/ml	~30 μl
H_2O	—	*qs* to 30 ml

[a] Formamide should be deionized before using: Place 500 ml of formamide in a 50-g bed of 20 to 50 mesh mixed bed resin (available from Bio-Rad). Stir for 1 hour and filter. Aliquot in 50-ml tubes and freeze at –20°C.

[b] Denature the cRNA probe and salmon testis DNA by boiling for 15 minutes and cooling rapidly on ice for 5 minutes. Use immediately.

2. Prehybridization: Combine the hybridization solution components (excluding the ^{32}P probe) into solution. Place 10 ml in a heat-sealing cellophane bag with one nylon blot. Remove as many bubbles from the bag as possible. Seal the bag and incubate at 42°C for at least 3 to 4 hours or preferably overnight, with mild shaking.
3. Hybridization: Cut open the cellophane bag and discard all of the solution. Prepare 10 ml of the hybridization solution, this time with the ^{32}P probe. Make sure the salmon DNA and the ^{32}P probe have been appropriately denatured as described earlier. Reseal the bag and let this incubate at 42°C for 48 hours. Cut open the bag and pour all liquid into a radioactive waste container.

C. Posthybridization

All washings take place at 62°C in a shaking water bath.

1. First wash solution:

2× SSC	100 ml of 20× SSC stock
0.1% SDS	5 ml of 20% SDS stock
H_2O	*qs* to 1000 ml

   Wash with 500 ml for 15 minutes. Repeat.

2. Second wash solution:

0.2× SSC	10 ml of 20× SSC stock
0.2% SDS	10 ml of 20% SDS stock
H_2O	*qs* to 1000 ml

   Wash with 500 ml for 15 minutes. Repeat.

3. Third wash solution:

0.2× SSC	10 ml of 20× SSC stock
0.4% SDS	20 ml of 20% SDS stock
H_2O	*qs* to 1000 ml

Wash with 500 ml for 15 minutes. Repeat.

The results of the hybridization are visualized either by exposing the membrane to X-ray film overnight or by using a specialized scintillation counter such as a Betascope (Betagen Corp.) or Packard Instant Imager. These instruments are expensive, but they provide quantitative information about the amount of radioactivity bound to the membrane and also provide an image of the hybridization comparable to an autoradiograph.

If this equipment is not available, autoradiography should be carried out by sandwiching X-ray film between the membrane and a phosphorescent intensifying screen in a light-tight cassette and carrying out the exposure at –70°C. The cassette, intensifying screens, and X-ray film are available from Kodak and other companies, as are the darkroom supplies needed to develop the film. Different types of film are available. Faster films generally are preferable as they permit shorter exposure times, but they have a larger grain-size and their resolution may be limited.

VIII. Slot Blotting

Northern blotting is useful and necessary for demonstrating that the probe being used is specific for a single mRNA species. However, it is too time consuming to be used for quantitating the amount of specific mRNAs in a large number of samples, such as might be needed to demonstrate a change in gene expression following teratogenic insult. Therefore, once you have assured yourself through Northern blotting that your probe is specific, slot blotting is a more economical way of processing a large number of samples. Slot blots are similar to Northern blots in that RNA is fixed to a nylon or nitrocellulose membrane and hybridized to a cRNA or cDNA probe. The difference is that the RNA from one's sample is pipetted directly onto the membrane without being electrophoretically separated. Commercially available slot blot manifolds allow the loading of a large number of samples (72, typically) on a single membrane. The wells in the manifold in which the samples are loaded are oblong, or slot-shaped, hence the name slot blots. Once the samples are loaded, the membrane is treated exactly like a Northern blot. Slot blots make it possible to load several concentrations of the same sample of RNA onto a single membrane, thereby permitting one to determine the concentration range that produces a linear response in terms of amount of radioactive probe hybridized. This is an important feature when one is trying to quantify changes in mRNA expression.

IX. RNase Protection Assay

The RNase protection assay is another technique for detecting and quantitating a specific mRNA. In this assay, the sample RNA and the cRNA probe are hybridized in solution and then treated with ribonuclease, which degrades all single-stranded RNA but leaves the cRNA-mRNA complex intact. The complex is then analyzed on polyacrylamide gel. This method is reasonably easy and sensitive. The cRNA-mRNA complex can be cut from the gel and counted in a gamma counter, making the assay quantitative.

Procedure

1. Make the probe according to the instructions for the kit specific to the vector containing the cDNA of interest.
2. Precipitate the sample RNA in ethanol as described above. Reconstitute about 10 μg RNA in hybridization buffer containing 5×10^5 cpm of probe RNA. For the hybridization buffer, make a 5× stock of 200 m*M* PIPES (pH 6.4), 2 m*M* NaCl, and 5 m*M* EDTA. Dilute to 1× with deionized formamide on the day of use.
3. Incubate at 85°C for 10 minutes.
4. Incubate overnight at 45°C (a higher temperature is necessary for some probes, probably to overcome conformational changes that may occur idiosyncratically).
5. Add 350 μl of a solution of 40 μg/ml RNase A and 2 μg/ml RNase T1. Incubate for 60 minutes at 30°C.
6. Add 10 μl of 20% sodium dodecyl sulfate and 2.5 μl of 20 mg/ml proteinase K. Incubate at 37°C for 20 minutes.
7. Extract with an equal volume of salt-saturated phenol/chloroform/isoamyl alcohol (24/24/1). Remove the upper phase to a clean microcentrifuge tube containing 1 μl of yeast tRNA (10 mg/ml).
8. Precipitate with 2.5 volumes ethanol. Redissolve the pellet in 5 μl of a loading buffer consisting of 80% formamide, 1 m*M* EDTA, 0.1% bromphenol blue, and 0.1% xylene cyanol.
9. Heat at 85°C for 5 minutes.
10. Load onto a 6% polyacrylamide gel (29:1, acrylamide:bisacrylamide), 0.4 mm thick and 14 to 20 cm long. Run at 300 V until the bromphenol blue (faster running dye) is 75% down the gel. Note that this gel is nondenaturing, but the RNA is denatured by heating and formamide.
11. Remove the gel from the glass plates, wrap in cellophane, and expose to X-ray film or analyze on an imaging system as described for blots. Usually, it is also possible to locate the radioactive bands on the gels by autoradiography and to cut them out and count them in a scintillation counter to quantitate binding. Figure 1.3 shows typical results of an autoradiograph from an RNase protection assay.

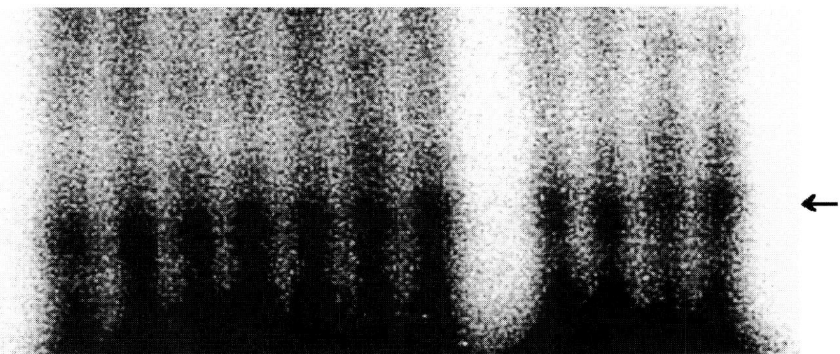

FIGURE 1.3
Results of an RNase protection assay. The image is an electronic autoradiograph of the gel. The band indicated by the arrow is MT-1 mRNA hybridized to a cRNA probe. The heavily stained area at the bottom of the gel represents the degraded RNA and cRNA probe resulting from RNase treatment.

X. Use of Oligonucleotides

Oligonucleotides are small sections of DNA that can be synthesized in any desired sequence. If one knows the sequence of the mRNA to be detected, it is possible to have a complementary oligonucleotide synthesized at low cost. Most institutions that have molecular biology programs have a DNA synthesizer. For those that do not have access to this equipment, a large number of companies offering oligonucleotide synthesis place classified advertisements in *Science* each week. The oligonucleotides are made radioactive by a 5′-end labeling procedure and can be used for Northern and slot blots instead of a cRNA probe.

XI. Other Resources

This chapter provides only a brief introduction to molecular biology techniques. There are many encyclopedic molecular biology lab manuals that you should consider obtaining if molecular approaches become routine in your laboratory. Two of the best are

Current Protocols in Molecular Biology, edited by F.M. Ausubel, R. Brent, R.E. Kingston, D.D. Moore, J.A. Smith, J.G. Seidman, and K. Struhl, Wiley Interscience, New York, 1989.

Molecular Cloning: A Laboratory Manual, 2nd ed., by J. Sambrook, E.F. Fritsch, and T. Maniatis, Cold Spring Harbor Laboratory Press, Plainview, NY, 1989.

Appendix A: Reagents Used in Molecular Biology

1. **Denhardt's Solution:** Make up a 100× stock solution, which is filtered and stored as aliquots at –20°C. Dissolve 10 g Ficoll, 10 g polyvinlypyrrolidone, and 10 g BSA (Fraction V), in 500 ml of H_2O.
2. **DEPC-Treated Water** (not for Tris buffer):
   a. Dissolve 1 ml of diethylpyrocarbonate (DEPC) in 1 l of distilled water.
   b. Let stand overnight, then autoclave.
3. **LB (Luria-Bertani) Broth:**
   a. Dissolve in 950 ml distilled water:
      10 g bacto-tryptone
      5 g bacto-yeast extract
      10 g NaCl
   b. Adjust to pH 7.0 with 5 *N* NaOH.
   c. *qs* to 1 l with distilled water.
   d. Autoclave.
4. **LB Agar:**
   a. Add 15 g of bacto-agar to 1 l of LB broth just prior to autoclaving. Mix carefully after removal from the autoclave.
   b. Wait until the solution has cooled to 50°C before adding ampicillin or other antibiotic.
5. **Salt-Saturated Phenol:**
   a. Melt a 1-lb bottle of high-grade phenol at 68°C.
   b. Add an equal volume of 0.5 *M* Tris Cl (pH 8.0).
   c. Stir for 15 minutes, then let stand until two phases appear. Discard the upper aqueous phase.
   d. Add an equal volume of 0.1 *M* Tris (pH 8.0) to the phenol. Stir for 15 minutes. Let the phases separate as above and discard the upper phase.
   e. Check the pH of the phenol using pH paper. If it is lower than pH 7.8, repeat step d, which can be repeated as many times as necessary to raise the pH to ≥ 7.8.
   f. Add 0.1 volume of 0.1 *M* Tris (pH 8.0) containing 0.2% 2-mercaptoethanol to the equilibrated phenol. This will form an aqueous layer on top of the phenol and retard oxidation.
   g. Salt-saturated phenol may be stored up to 1 month at 4°C in a dark bottle.
6. **SSC Buffer:** This can be made as a 20× solution containing:
   175.3 g NaCl
   88.2 g Na_3 citrate-2 H_2O
   Dissolve in 800 ml water. Adjust pH to 7.0 using 1 N HCl. *qs* to 1 l.
7. **SSPE (20×):** Dissolve 175.3 g of NaCl, 27.6 g of NaH_2PO_4 H_2O, and 7.4 g of Disodium Salt EDTA in 800 ml of H_2O. Adjust the pH to 7.4 with NaOH (need about 6.5 ml of 10 *N* solution). Adjust the volume to 1 l with H_2O. Dispense into aliquots. Sterilize by autoclaving.

I. Introduction

The identification of the expression pattern of genes in embryonic and adult tissues provides critical information about the temporal and spatial action of genes and thus constitutes a first important step in understanding their function. The most straightforward method to visualize gene expression patterns in space and time is *in situ* hybridization on tissue sections or on whole mount specimens. The sole prerequisite for carrying out *in situ* hybridization is having tissue preserved by methods that maintain the integrity of mRNA and having a fragment of expressed sequence. Such fragments can be prepared readily by subcloning of previously isolated cDNA clones or by reverse transcriptase-polymerase chain reaction (RT-PCR) methods.[1,2]

The principle underlying *in situ* hybridization is to anneal a labeled DNA or RNA probe to the cellular RNA, made accessible either by sectioning of the tissue of interest or by making whole mounts permeable to the probes.[3-5] *In situ* hybridization is very sensitive, particularly when antisense RNA probes are used. The procedure consists of four steps: (1) The preparation of tissue (fixation, dehydration, embedding, and sectioning), (2) the preparation of the probe, (3) the application and annealing of the probe and subsequent removal of the nonhybridized probe by RNase digestion and washing, and (4) the visualization of the expression pattern (autoradiography, color reaction).

Riboprobes used in *in situ* hybridization experiments are either labeled with ^{35}[S] uridine-5′-triphosphate (UTP) or with a hapten (most frequently digoxygenin). In the first case, hybridization is detected by autoradiography, while in the second case antibodies are used to visualize the hapten. The decision of whether to determine a pattern of expression on sections or on whole mounts is largely determined by the nature of the specimen. Larger pieces of tissues or older embryos usually are not suitable for the whole-mount procedure. Once an epidermis covers the embryo, it becomes impermeable for most reagents, including riboprobes; however, even larger specimens are fine for the whole-mount procedure if the gene of interest is expressed on the surface of the specimen. Larger specimens can be dissected into parts, and those may be sufficiently permeable for the whole-mount approach. For whole mounts, digoxygenin-labeled probes are used, as autoradiography required for the detection of radiolabeled probes is not applicable to three-dimensional objects. Also note that whole mounts subjected to hybridization with a digoxygenin probe can be sectioned and examined under Nomarski optics. *In situ* hybridization on sections using ^{35}[S] UTP-labeled or digoxygenin-labeled probes is a method ideal for larger specimens such as the brain. While offering less contrast than radiolabeled probes, digoxygenin probes can provide cellular resolution. It goes without saying that whole mounts and section approaches are complementary, and both may be required for a detailed expression analysis of a developmentally regulated gene.

Essential to all *in situ* hybridization analyses is the use of controls that demonstrate probe specificity. A commonly accepted control is the use of sense riboprobes. The hybridization of antisense and sense riboprobes in parallel experiments and under identical conditions is particularly important if a gene appears to be broadly expressed. Once hybridization and washing conditions are established, hybridization with the sense control may be omitted. However, when a new tissue or significantly different embryonic stage is examined, then the use of sense probe controls is strongly recommended. It is difficult to determine which portion of a cDNA is most suitable for hybridization. If the gene of interest is a member of a gene family, we advise the use of a riboprobe complementary to the 3′ untranslated region, which usually is not conserved. However, a riboprobe designed to recognize the protein coding region gives good results in most cases, since the protocols described here use an RNase digestion step followed by a stringency wash. Even a few mismatches are sufficient for RNase to cleave the probe leading to the removal of the cleaved probe fragments during the stringency wash.

This chapter is designed to discuss each of the steps in sufficient detail in order to enable the reader to carry out *in situ* hybridization. However, we assume that the reader is sufficiently familiar with basic molecular biology techniques (subcloning, restriction analysis, electrophoresis etc.) and thus do not describe such fundamentals (for more information, see Reference 6). The protocols we chose were successfully used in our laboratory using embryos and tissues from mouse,[7,8] chicken,[9] and human.[7,10]

II. Tissue Preparation

A. Fixation

The general purpose of fixation is to preserve good histological detail. Because the strength of the hybridization signal is dependent upon the availability of cellular mRNA that has been preserved in the tissue throughout the procedure, fixing protocols that cross-link excessively are not suitable. Glutaraldehyde and formaldehyde are particularly suitable for *in situ* hybridization protocols. It is important to keep in mind that not all tissues require the same amount of fixation and that steps in the protocol are interdependent. For example, shorter fixation times or the use of gentle fixatives usually require a shorter protease digestion. A good and popular fixative for most embryonic and adult tissues is 4% paraformaldehyde in phosphate buffered saline (PBS). Fixations usually are performed at 4°C, with gentle agitation. All dissections are performed in sterile, ice-cold PBS. The tissue is rinsed in fresh PBS after dissection is complete, and placed into the chosen fixative for approximately 12 hours. Convenient containers for fixation include glass scintillation vials (20-ml volume) or, if smaller tissues are being fixed, 24-well plates. While fixations of larger tissue pieces can be extended to days, care should be taken not to fix an excessive amount of tissue in a single vial. A useful guideline is that the volume of the fixative should be ≥30 times that of the tissue to be fixed.

Alternative fixation agents include Bouin's or Carnoy's reagent, both of which have been used successfully in our laboratory for *in situ* hybridization of sections. Bouin's fixative is particularly useful when calcified tissues need to be sectioned (older embryos), since this fixative softens hard tissues. In the case of Carnoy's and Bouin's fixative, contents in the amniotic fluid become very solid and are hard to remove after fixation. Thus, embryos should be washed in saline and extra embryonic membranes should be removed before fixation. PFA and MEMFA are equally suitable to fix embryos for whole mount studies (see below).

Recipes

MEMFA

- 0.1 *M* MOPS (3-[*N*-morpholino]propanesulfonic acid), pH 7.4
- 2 m*M* EGTA
- 1 m*M* $MgSO_4$
- 3.7% formaldehyde
- A stock solution is prepared not containing the formaldehyde.
- 1 part of a 37% formaldehyde stock solution is added to 9 parts of buffer/salt mix just prior to use.

4% Paraformaldehyde/PBS

- 100 ml of paraformaldehyde (PFA) fixative is prepared by heating 90 ml H_2O and 100 μl 2 *M* NaOH in a microwave oven to 50°C.
- The solution must be basic to aid in the dissolution of the PFA.
- Add 4 g of PFA (use a face mask) and stir in a hood until the PFA is dissolved.
- Add 10 ml 10× PBS, filter through a Whatman paper filter, and let cool to room temperature.

- If necessary, adjust the pH back to 7.4 by the addition of 100 µl 2 *M* HCl.
- If the PFA is too acidic, it will destroy DNA; if it is too basic it will destroy RNA.
- Store the solution at 4°C, sealed with parafilm to prevent fixation of other proteins stored in the refrigerator. If kept refrigerated, the solution is usually good for about one week.

Carnoy's Fixative

- 100% EtOH (60 ml)
- Chloroform (30 ml)
- Acetic acid (10 ml)

Bouin's Fixative

- Saturated picric acid aqueous (75 ml)
- Formalin (25 ml)
- Acetic acid (5 ml)
- Suitable for larger embryos since this fixative decalcifies the ossified tissue.
- To fix mouse embryos older than 17 dpc, either the belly has to be opened up or the tissue of interest should be isolated by dissection from the rest of the body for better penetration of fixative.
- The skin of the embryo from this stage onwards is impermeable to the fixative.

B. Dehydration

During dehydration, water is removed from the specimen so it can be embedded in paraffin. Moreover, in a dehydrated state, specimens can be stored at –20°C for months without noticeable RNA degradation. Embryos are taken through a graded ethanol/salt series, beginning with physiologic salt concentration and ending in 100% ethanol (EtOH). The length of each wash is dependent upon the specimen size. Stage 6 through 14 chick embryos, or mouse embryos younger than day 9, are washed for 10 minutes per step. In the case of a stage-20 chick embryo or a day-11 mouse embryo, 15 to 30 minutes are sufficient; day-16 mouse embryos are washed for 60 to 90 minutes in each step. Very large tissues should be dehydrated for 2 to 4 hours per step, with the 70% EtOH step being extended overnight. This may need to be adjusted if the tissue is less permeable (i.e., from an older embryo or an adult). Transferring the specimen through the dehydration steps too rapidly results in specimens that cannot be sectioned. If water is not completely replaced by ethanol, or if the wax is too hot or the wax permeation period too short (Sectio III.A), air bubbles develop and appear as holes in the paraffin sections which results in tearing of the section.

To dehydrate, subject tissue to the following steps at 4°C on a rocking platform moving at a gentle rate (20 times per minute):

1. 0.9% NaCl
2. 30% EtOH in 0.9% NaCl
3. 50% EtOH in 0.9% NaCl
4. 70% EtOH in H_2O
5. 90% EtOH in H_2O
6. 100% EtOH

At this point, embryos can be stored indefinitely at –20°C.

To dehydrate with Carnoy's or Bouin's fixatives, Carnoy's fixed tissues are transferred to 100% EtOH for 15 minutes, 4°C (twice), and are stored in 100% EtOH at –20°C. Bouin's fixed embryos are transferred to 70% EtOH for 15 minutes, 4°C. Wash in 70% EtOH and replace until the yellow picric acid is gone; otherwise, this agent crystallizes in the embryo. Transfer specimen to 95% EtOH and then to 100% EtOH for 15 minutes, 4°C (each twice), and store in 100% EtOH at –20°C.

Specimens to be used for whole mounts fixed in PFA are dehydrated as described above (steps 1 to 6) except that methanol is used instead of EtOH. If specimens were fixed in MEMFA, wash with 0.9% NaCl (2 to 5 minutes), transfer to 90% methanol, and store at –20°C.

III. *In Situ* Hybridization on Sections Using Radioactive Probes

A. Embedding

Paraffin wax should be melted ahead of time, either by placing a container in a warming oven overnight or by microwaving the wax for approximately 10 minutes and leaving a small amount of wax unmelted. Care should be taken not to overheat the wax, as this destroys its ability to form the proper lattice structure when the wax cools. Discard unused wax after 2 to 3 days. A small aliquot of the wax should be placed in a separate beaker, and, approximately 10 minutes before use, an equal amount of xylene is added. Discard the unused portion of the xylene:wax solution immediately in a receptacle in a hood (xylene is toxic).

Prior to embedding, process the tissues through the following solutions on a rocking plate, using the same length of time as was employed when transferring tissue from 0.9% NaCl to ethanol:

1. 100% EtOH, at room temperature
2. EtOH/xylene 1:1, at room temperature

3. Xylene, at room temperature
4. Xylene/wax, at 58°C (does not need to be rocked)
5. Wax, 3 times at 58°C

After the final wax wash, prepare a suitable container for the tissue, such as a small weigh boat or a commercially available plastic mold. Gently swirl the vial to suspend the tissue in the melted wax and quickly pour it into the container; avoid creating bubbles in the process. The tissue can be oriented using forceps that have been warmed over an alcohol flame. If the tissue fails to pour from the vial, add more wax and try again. Bubbles in the molten wax can be removed by heating the forceps and gently moving it through the wax. Bubbles that are in close proximity to the tissue should be removed, as they will create voids in the wax that will make sectioning more difficult. The wax blocks should be allowed to cool at room temperature without disruption and can be stored at room temperature indefinitely. Do not refrigerate or freeze wax blocks.

Tissue fixed with Bouin's fixative and stored in 100% EtOH is embedded as follows. Transfer specimen to methylbenzoate with three exchanges at room temperature; after the last change, wait until the specimen sinks to the bottom of the vial. Transfer to benzene (twice, room temperature), benzene and wax (1:1, 58°C), wax (twice, 58°C) and pour into mold. Methylbenzoate and benzene are used instead of xylene, because the latter hardens the tissue and makes it too brittle for sectioning.

Hints and Problems

1. To aid in positioning the tissues in wax, stain tissues with Eosin Y dye in EtOH prior to embedding. For a 1% stock of alcoholic eosine, 1 g of Eosin Y is dissolved in 20 ml of water and 80 ml of 95% EtOH are added. Prior to use, mix 1 part stock solution with 3 parts of 80% EtOH.
2. Minimizing the length of the xylene step is advisable, as extended exposures cause tissue to become brittle.
3. Overheated wax shrinks after it has been poured, creating a depression in the wax block. Do not attempt to re-pour wax into this depression as it will not adhere to the original wax, causing separation when the wax block is mounted for sectioning. Instead, the entire wax block is melted by placing it back into the oven for the minimum time necessary to completely melt the wax. Re-pour and let cool to room temperature.

B. Sectioning

Mounting the tissue onto a wooden block (or an equivalent support) is easily performed if there is sufficient wax around all sides of the embedded tissue. Determine the plane of section that is desired, then spread a small amount of molten wax onto the block and heat up a metal spatula until it is very hot. Place the warmed spatula against the wax on the block and simultaneously lower the tissue-containing wax block towards the top surface of the heated spatula. Quickly remove the spatula and firmly place the now-melted surface of the wax block into the heated wax on the wooden block. Reheat the spatula and melt the four sides of the wax block slightly, allowing the dripping wax to fill in any space between the wooden block and the tissue-containing wax block; allow sample to cool at room temperature.

In order to produce a ribbon of tissue sections that is straight and wrinkle free, the wax block must have 90° angles. If the bottom edge is uneven, or either side is longer than the other, the ribbon will twist and make it difficult to position the sections onto a slide. Gradually shave down the wax block around the tissue with a clean razor blade, preserving enough wax at the periphery so that the tissue occupies approximately one half of the total wax surface. Take care to remove only small amounts of wax at a time so that the block does not crack.

Sections are best cut in a place with few air currents and constant temperature. Position a water bath (set between 45 and 50°C and use autoclaved water) and a slide warmer set at 38°C in close proximity. The position and angle of the microtome blade, the speed at which to cut, and the optimal thickness of the section are determined empirically. To begin, use a blade angle of 10°, 7-μm thickness, and a slow cutting speed. If the sections tend to wrinkle, it may be due to compression as a result of cutting too fast; thus, slow down when passing through the tissue. In addition, one can also reduce the angle of the knife. If the sections fail to form a ribbon and come off as individual sections, it is frequently due to the angle of the blade; adjust blade to a larger angle. If you detect a scratch or a split in the wax section, first check that there are no obvious nicks in the blade. If so, reposition the microtome blade so that a new zone is being used; use extreme caution when doing this so as not to cut yourself. Dust particles in the block underneath the specimen also can cause failure to form a ribbon. Shave a small amount of wax off of the bottom edge of the wax block (the edge where the knife first contacts the block).

Once you have succeeded in making ribbons of sections, lay them onto a clean piece of paper. It is particularly helpful to use black paper since you can see the tissue in the wax with more contrast. After you have accumulated a number of ribbons, you can begin to mount them. Cut the ribbons into sections that are shorter than the slide you are using; the ribbon will expand approximately 10% as it warms up.

Place the ribbon gently into the waterbath, taking care to place the side of the ribbon that faced the blade in contact with the water (shiny side of ribbon). Watch the ribbon closely; as the wax melts and expands, small wrinkles should disappear. Do not allow the water to get too warm, as the wax will begin to melt and in doing so will tear the tissue. To mount the tissue, position a slide underneath the ribbons and, using forceps to guide the sections, slowly lift the slide out of the water. Care should be taken not to touch the tissue since the wax will melt slightly and become adherent to the forceps. Allow the slides to dry on the slide warmer for a minimum of 2 hours, then store them in a cool (4 to 20°C), dry place. Include a desiccant in the box, then seal the box. Avoiding water condensation is critical; in the presence of water, RNase can start cleaving the cellular RNA.

Recipes

There are a number of protocols for treating slides to aid in adherence of the wax sections. We present two methods, one of which cleans and etches the slides slightly using TESPA (3-aminopropyltrioxyethyl silane); these slides are stable indefinitely. The other method uses poly-L-lysine; although the tissue adheres very well, poly-L-lysine is labile and the slides are more difficult to prepare.

TESPA

- Place slides in a stainless steel rack and dip in 10% HCl/70% EtOH for 2 minutes, followed by distilled water and then 95% EtOH, 2 minutes each.
- Dry the slides either in a baking oven at 150°C for 15 to 20 minutes or in a speed-vac, then cool to room temperature.
- Dip slides for 10 to 15 seconds in 2% TESPA in acetone, followed by two quick washes in acetone and one in water.
- Dry at 42°C and place in a vacuum or desiccated box.

Poly-L-lysine

Place slides in a stainless steel rack and wash in a 1% solution of Linbro ("7X") (Flow Labs, Inc.) for 15 minutes. Rinse the slides extensively in deionized water (5 times, 2 to 3 minutes each), then bake them at 150°C for 1 hour. Cool to room temperature before immersing the slides in a freshly prepared solution of 100 μg/ml poly-L-lysine in autoclaved deionized water for 1 hour at room temperature. Allow the rack to drain and air dry for a few hours. Store in a vacuum or desiccated box for up to 2 weeks.

C. Preparation and Synthesis of the Riboprobe

Single-stranded RNA probes are prepared using plasmid vectors containing a polylinker bordered by promoters of T3 and T7 bacteriophage RNA polymerases. Probe synthesis reactions are carried out using a linearized plasmid as a template. The plasmid is linearized by digesting with a restriction enzyme that cuts at a single site downstream of the insert. The site should be oriented so that the synthesized RNA probe will be terminated as a result of polymerase run-off. In order to synthesize a probe that hybridizes to mRNA, the antisense strand of the cloned insert must be used as the template. Following restriction enzyme digestion, the template is treated with proteinase K, then extracted in phenol/chloroform and precipitated in ethanol. This proteinase K treatment removes ribonucleases from the template reaction. The linearized template is then resuspended in diethyl pyrocarbonate (DEPC)-treated water or TE buffer (10 m*M* TrisCl, 1 m*M* EDTA, pH 7.4).

Template Preparation

cDNAs should be digested to yield both sense and antisense orientations of the template. The sense probe serves as an important control for nonspecific hybridization. At least 10 μg of DNA should be digested initially. Determine that the restriction digestion has gone to completion by running a small aliquot of the reaction out on an agarose gel; the presence of the uncleaved template will generate RNA transcripts that contain polylinker and vector sequence and may be a source of background signal. Add 50 μg/ml of proteinase K to the restriction buffer and incubate for 30 minutes at 37°C, followed by two phenol/chloroform (1:1 v/v) extractions and ethanol precipitation. Resuspend the digested, proteinase K-treated cDNA in TE made with DEPC-treated water.

Transcription Reaction

A typical transcription reaction is described that will generate high-specific activity probes (transcription kits are commercially available). Add the solutions in the following order, keeping the reagents on ice and using caution to avoid contamination with RNases (gloves, autoclaved tubes and tips, etc.).

- 6 μl of 5× transcription buffer (5× is 200 m*M* Tris-HCl pH 8.0, 40 m*M* $MgCl_2$, 250 m*M* NaCl, 10 m*M* spermidine)
- 1 μl of 1 μg/μl restricted, proteinase K-treated, linearized DNA template
- Add sufficient DEPC-treated water to make a final volume of 30 μl
- 1 μl of 10 m*M* rATP
- 1 μl of 10 m*M* rCTP
- 1 μl of 10 m*M* rGTP
- 1 μl of RNAsin (40 U/μl)
- 1 μl of 0.75 *M* DTT (use DTT that has been thawed once)
- 5 to 10 μl of 400 to 800 Ci/mmol, 10 μCi/μl ^{35}[S] UTP
- Mix well, spin, and then add 10 U of T3 or T7 RNA polymerase. Incubate at 37°C for 2 to 3 hours.

To remove the template DNA, use DNase I that is RNase free, and add the following to the 30-μl transcription reaction: 19 μl of DEPC-treated water, 1.7 μl of 0.3 *M* $MgCl_2$, and 2 U of DNase I; incubate at 37°C for 15 minutes.

The probe is precipitated by adding the following to the transcription reaction:

- 100 μl of DEPC-treated water
- 100 μl of 1 mg/ml yeast tRNA (previously purified by phenol extraction)
- 250 μl of 4 *M* ammonium acetate
- 1 μl of EtOH

Vortex well, precipitate 5 minutes on ice, spin 10 minutes, and remove supernatant. This precipitation is repeated once. Aspirate off the remaining supernatant using a pulled-out Pasteur pipet. A speed-vac also can be used for this purpose; however, use caution, as RNA pellets may be difficult to resuspend if overdried. Resuspend pellet in hybridization buffer (see Section III.D.2). Alternatively, unincorporated nucleotides are removed by passing the transcription reaction over a column containing RNase-free Sephadex G-50.

If hydrolysis is necessary to reduce the size of the transcripts (longer than 1 to 1.5 kb, while the optimal probe size is 0.5 to 0.8 kb), dissolve the probe in water and add an equal volume of 80 m*M* $NaHCO_3$, 120 m*M* Na_2CO_3. Incubate at 60°C for t minutes, $t = (L_o - L_d)/0.11\ L_o L$, where L_o is the original probe length in kilobases and L_d is the desired probe length.

Determination of Incorporation

Typically, greater than 50% of the input ^{35}[S] UTP is incorporated into the RNA transcription reaction. The use of labeled UTP as the single source of this nucleotide limits the mass yield of RNA probe. The following calculation is used to determine the amount of the RNA generated in the transcription reaction:

$$\frac{\text{Total number of } \mu\text{Ci added} \times 13.2 \times \%\text{ incorporation}}{\text{Number of } ^{35}\text{[S]dNTP used} \times \text{specific activity of radioisotope}}$$

$$= \text{ng of RNA synthesized}$$

Percent incorporation is determined by total counts divided by incorporated counts. Count an aliquot of riboprobe and an aliquot of the reaction mix taken prior to precipitation in 2 ml water-miscible scintillation fluid in the ^{14}C channel. The approximate specific activity of freshly prepared RNA is expected to be 10^9 cpm/µg of RNA. If necessary, store the probe at -80°C for up to a week (fresh probes are better because older probes display strand breaks).

D. Hybridization Steps

1. Dewaxing and Fixation of Sections

All solutions used for prehybridization should be RNase free. This is achieved by autoclaving for 45 minutes. It is convenient to prepare 10× stock solutions and make dilutions as required. Most of the prehybridization solutions can be reused if they are kept RNase-free. Paraformaldehyde (PFA) solutions can be re-used twice but should be discarded thereafter. PFA is stored at 4°C. Proteinase K and acetylation solutions must be made fresh each time. We use a Tissue-Tek II staining station from Miles for the steps described here. It is advisable to cover all solutions containing ethanol with plastic wrap to prevent evaporation. Place slides (in a rack) in the following solutions in the order listed:

1. Histo-Clear (National Diagnostics) or another xylene substitute to remove wax, 10 minutes, twice.
2. Ethanol series, starting at 100% and progressing to 95, 70, 50, and 30%. Agitate the slides in each solution until equilibrated (approximately 1 minute each).
3. 0.9% NaCl, 5 minutes
4. 1× PBS, 5 minutes
5. 4% PFA, 20 minutes (this solution can be re-used below)
6. 1× PBS, 5 minutes
7. Subject slides to proteinase K treatment: 20 μg/ml proteinase K in 50 m*M* Tris-HCl pH 7.6, 5 m*M* EDTA, 5 minutes. After this step, the tissue is particularly susceptible to RNases; use extreme caution.
8. 0.2 *M* HCl, 5 minutes (can be left out)
9. 1× PBS, 5 minutes
10. 4% PFA, 20 minutes
11. Acetylation: The purpose of this step is to acetylate amino groups in tissues to reduce electrostatic binding of probe.[12] Acetylation also blocks binding of probe to poly-L-lysine treated slides. These steps must be performed in a hood or another well ventilated place. Suspend the slide rack above a rapidly rotating stir-bar in a solution of 0.1 *M* triethanolamine-HCl (TEA; pH 8.0). Add 600 μl of acetic anhydride; wait 5 minutes, then add 600 μl more of acetic anhydride. Incubate for a total of 10 minutes. Care should be taken not to introduce water into the acetic anhydride stock; acetic anhydride would rapidly hydrolyze to acetic acid in the presence of water. Thus, if the agent smells like acetic acid, it should be discarded.
12. 1× PBS, 5 minutes
13. 0.9% NaCl, 5 minutes
14. 30% and 50% EtOH, 1 minute each
15. 70% EtOH, 5 minutes (this step is longer to make sure to remove all salt deposits)
16. 80, 95, 100, and 100% EtOH, 1 minute each

Air-dry slides in a dust-free, RNase-free place, then store in a sealed container at room temperature with desiccant for up to a few days. Theoretically, dewaxed slides can be stored indefinitely; however, they are much more susceptible to RNase once the tissue has been treated with proteinase K and are also prone to oxidation.

2. Prehybridization

Sections may be incubated with hybridization buffer not containing the probe in order to reduce a high background; this may help especially if the mRNA of interest is rare. For 25 × 75 mm slides with 24 × 50 mm cover slips, 100 μl of hybridization mix per slide is sufficient. Be sure to prepare extra mix, because the viscosity of the hybridization mix leads to large pipetting errors. To the mix, add 1/100 th volume of 25 m*M* alpha-S-thio ATP (this is crucial for the reduction of background). The mix is then heated in 90 to 100°C water for 2 minutes and cooled to room temperature. Carefully place the hybridization mix onto the dry slide, then lower a cover slip onto the slide with forceps. This is easiest if you drop one end of the cover slip onto the slide, then slowly lower the rest. Try to avoid bubbles, but if they do form, make sure that hybridization mix completely covers every section. Incubate slides in horizontal position for 1 hour at 50°C in a chamber, the bottom of which is filled with several layers of Whatman paper soaked in 50% formamide/2× SSC. We have used chambers consisting of transparent plastic obtained in hardware or department stores. We place the slides on a grid rack built from plastic pipets.

Prehybridization has disadvantages. First, it necessitates removing the cover slip covering the section which may damage the tissue. Second, the exact probe concentration cannot be determined for the subsequent hybridization, as hybridization buffer remains on the slide and will dilute the probe. The alternative, a hybridization performed at a higher temperature, can also decrease background and has the advantage of being quicker and not requiring the removal of the cover slip.

Recipe

Hybridization buffer

- 50% deionized formamide
- 0.3 *M* NaCl, 20 m*M* Tris-HCl (pH 8)
- 5 m*M* EDTA
- 10% dextran sulfate (heat gently and rock to dissolve)
- 0.02% Ficoll
- 0.02% bovine serum albumin (RNAse free)
- 0.02% polyvinylpyrrolidone
- 0.5 mg/ml yeast RNA (RNase free)
- Make 1 ml aliquots and store at –80°C until use.

3. Hybridization

An optimal signal-to-noise ratio is achieved when probe concentrations are just sufficient to saturate the complementary cellular mRNAs. However, this concentration is usually unknown, and the optimal probe concentration is therefore determined empirically. Initially, the amount of probe used in a hybridization reaction should be in tenfold or greater excess of the transcripts of interest. A good starting place is to use a high specific activity probe (10^9 cpm/µg riboprobe) at a concentration of 2 to 6 × 10^6 cpm per slide (2 to 6 ng). We have found, however, that using half or even less (1/10th) of that amount can yield even nicer *in situs* when the expression of the gene of interest is confined to a small area. The correlation is not intuitive; it seems that the lower the expression level of the mRNA species, the lower the optimum concentration of the probe. It is probably a good idea to start off with the 1 ng per slide and then to subsequently titrate to find the optimal concentration. The rate of hybridization depends on the concentration of probe. We found that overnight hybridizations are sufficient, although times as short as 5 hours have been used successfully.

If the sections were prehybridized, remove the cover slip by tilting the slide and simply letting it slide off. A 26 × 75-mm slide can be covered adequately with 100 μl of hybridization solution (50 μl are sufficient for prehybridized slides); calculate the appropriate amount of probe to add to the hybridization buffer and place into an Eppendorf tube. Place the tube into a 70°C waterbath for 3 minutes. It is not necessary to denature RNA probes prior to hybridization. However, heating the hybridization buffer prior to pipetting is helpful due to the viscous nature of the solution. Add 1/10 volume of 1 *M* DTT, vortex well, spin, and aliquot onto each slide. DTT is heat labile and therefore should not be added to the solution prior to heating. Apply the mix with a Pipetman and gently lower a cover slip onto the slide using forceps; decrease the angle between the cover slip and slide so that bubbles are forced out. The cover slips must be very clean to minimize bubbles. Do not press down on the cover slip to remove bubbles, as this will result in a thinner layer of the hybridization solution. Place the slides into the humidified, sealed chamber (see Section III.D.2). Incubation temperatures are determined empirically; a good starting point is 45 to 50°C, which is approximately 25°C below the T_m (melting temperature) of hybrids formed *in situ* with a 150-nucleotide long probe and a guanosine/cytidine (GC) content of 50%.

4. Posthybridization Washes

Washes are used to gently remove the cover slip, then to hydrolyze the nonspecifically bound probe which has adhered to the slide and the tissue. The optimal temperatures of the washes must be determined empirically; however, a temperature of 50°C will provide the same stringency as the hybridization conditions. Higher temperatures, up to 65°C, can be used to improve background without significant loss of signal. Proceed as follows:

1. Cover slip removal: Gently place the slides into a slide holder, using extreme caution not to dislodge the cover slips. Do not attempt to manually remove cover slips, as the shear force is likely to damage the tissue. Place the slide rack into a solution of 5× SSC at 50°C, add 40 m*M* beta-mercaptoethanol (in a hood!). Gently agitate the slides every few minutes; after 15 minutes the cover slips should have loosened sufficiently so that upon raising a slide out of the holder, the cover slip is left behind in the solution. Discard this solution in the radioactive waste.
2. Treat slides in 50% formamide, 2× SSC, and 40 m*M* beta-mercaptoethanol at 55 to 65°C for 30 minutes.
3. Wash slides once or twice in 0.5 *M* NaCl, 10 m*M* Tris-HCl (pH 8.0), 1 m*M* EDTA, and 40 m*M* beta-mercaptoethanol for 30 minutes at 37°C.

4. The nonspecifically bound probe is hydrolyzed using 10 to 20 μg/ml RNase A in 0.5 *M* NaCl, 10 m*M* Tris-HCl (pH 8.0), and 1 m*M* EDTA, for 30 minutes at 37°C. RNase A is stored as a stock solution of 10 mg/ml which can be refrozen. If background is a problem, the addition of RNase T1 (at 1 U/ml) is recommended.
5. Slides are re-equilibrated in 0.5 *M* NaCl, 10 m*M* Tris-HCl (pH 8.0), and 1 m*M* EDTA without RNase A for 15 minutes at room temperature, then subjected again to a high stringency wash (step 2 above) of 50% formamide/2× SSC, 40 m*M* beta-mercaptoethanol.
6. Wash once in 2× SSC for 15 minutes, then 0.1× SSC for 15 minutes, each at room temperature.
7. Dehydrate through an ethanol series: 30, 50, and 70% EtOH/0.3 *M* NH_4 acetate, then once in 95% and twice in 100% ethanol. Allow slides to dry. Step 2 (first stringency wash) and step 3 can be omitted if the probe behaves well (not "sticky").

E. Autoradiography and Visualization

To determine whether the stringency washes were of a sufficiently high temperature and to determine the length of time that slides should be exposed to emulsion, the slides are first exposed to a high performance autoradiographic film for 1 to 3 days. Cronex film (DuPont) is suitable for this purpose. Tape the top edges of the slides to an immobile surface (a used piece of film is ideal) and use care not to allow the slides to come in contact with dust or lint present in the film cassette. Ideally, antisense probe hybridization should reveal distinct hybridization in the sections, possibly only in specific regions such as the nervous system, the heart, or some other subset of tissues. The sense control should be essentially free of signal except for a weak uniform haze on top of the section. If this is not the case, the temperature of the stringency wash needs to be increased. Some genes are broadly expressed as would be revealed with the antisense probe. If this is suspected, it is extremely critical that the sense control be essentially blank. Sometimes there is intense signal around the margins of sections or even on the slide. This can be due to the slides drying out during hybridization. Other causes are discussed in Section III.F.

If it is determined that a higher wash temperature is needed to remove background, the slides can be rehydrated by placing them in 2× SSC for 15 minutes and then into a solution of 50% formamide/2× SSC/40 m*M* beta-mercaptoethanol. Rewashing can often remove nonspecifically bound probes; however, it rarely works as well as having washed initially at a higher temperature. After rewashing, dehydrate the slides as described under Section III.D.4, step 5.

1. Application of Emulsion

Slides of suitable quality are then dipped in photographic emulsion. Emulsion is prepared by adding 118 ml Kodak NTB-2 emulsion to 200 ml deionized water. Read directions carefully when using this emulsion. Most darkroom safelights are not suitable and will expose the emulsion; dark red (number 2) safelight is suitable. The emulsion and water must be warmed at 40 to 42°C for 30 to 35 minutes before mixing, using extreme caution to avoid exposure to any light source. It is convenient to aliquot the emulsion into glass scintillation vials, approximately 10 ml each. Wrap the scintillation vials in two layers of aluminum foil, then place inside a light-tight box. This box should be stored at 4°C and kept away from all forms of beta radiation.

Melt an aliquot of the emulsion at 42°C for 5 to 10 minutes (keep this time to a minimum since the emulsion tends to break down with heat and time). Do not agitate. Once melted, gently pour into a suitable container such as a slide mailer (e.g., Young Laboratories), using care not to create bubbles. Grasp the sides at the frosted end and gently lower them into the emulsion for 3 to 4 seconds; wipe the back of the slides to remove the emulsion and immediately place slides horizontally. The slides must be kept horizontal until the emulsion is completely hardened. This is dependent upon the airflow in the chamber in which the slides are drying. If a simple box is used, the edges usually require taping to exclude all light. A more convenient option is either to obtain a desiccant box which is airtight and place silica gel desiccant in it or to have a box manufactured. If a box is manufactured, it is convenient to have a fan installed. However, note that emulsion cannot be dried under vacuum, nor can it be heated as it tends to crack. Drying time is usually 5 to 7 hours at room temperature.

Once dry, the slides can be stored in a sealed black box containing silica gel desiccant. The box should be wrapped in aluminum foil, shielded from beta radiation and stored at 4°C. Use the exposure time of the Cronex film as a guide to determining the length of the emulsion exposure. Usually 3 to 4 times the length of the Cronex exposure is sufficient. Batches of emulsion should be tested prior to dipping tissue slides. Typically, the silver grain density should be so low as not to create a haze. Moreover, there should be no local clusters of silver grains.

2. Developing

Remove the slides from 4°C storage and let them equilibrate to room temperature. Using only the appropriate dark red (number 2) safelight, place the slides into a rack and develop as follows:

1. Kodak D-19 developer, 2 minutes (do not warm beyond room temperature)
2. Wash in water for 10 to 15 seconds
3. Place in fixer for 5 minutes
4. Final rinse in water for 10 to 15 minutes

After developing, the slides can be counter-stained with a solution of 2 μg/ml Höechst 33258 in water (stock solution of 10 mg/ml in DMSO; make 40 μl aliquots and store at –80°C). Dip the slides into the solution for 2 minutes, followed by 2 minutes in water. Air-dry the slides in a dark place, since the Höechst stain is light sensitive. Once dry, overlay the slides with 50 µl of a solution of 5 g Canada balsam in 10 ml methyl salicylate, then place a cover slip and wick off the excess mounting medium from the edges. Keep slides flat for 1 week so that the mounting medium dries out. Other mounting media may be used which dry faster; however, some of those exhibit autofluorescence and cannot be used in combination with Höechst staining. An alternative staining method not involving fluorescence detection involves staining with toluidine blue.[3]

3. Viewing and Photography

Höechst stain is a nuclear dye, which has the advantage of allowing a determination of cell density as well as visualizing tissue morphology. Using the 4′,6′-diamidino-2-phenylindole (DAPI) channel on a epifluorescence microscope at the same time that one views the silver grains by transmitted light allows one to determine the tissue which is expressing the gene of interest. Höechst dye eventually bleaches under fluorescent light; therefore, care should be taken to minimize exposure. To clean the slides for photography, remove residual emulsion with glacial acetic acid on a cotton swab. Remove acid with ethanol. One may also use a new razor blade and scratch off the emulsion on the back of the slide; do not scratch the glass itself since this will cause background in the darkfield. Typically, we obtain images of our *in situ* hybridization specimens by conventional photography. In this case, we take an image of the tissue revealed by blue Höechst fluorescence and superimpose a darkfield image. To enhance the contrast, we insert a red filter into the light path so that the silver grains become red instead of white. More convenient than classical photographs, the images are captured with a videocamera linked to a computer. These can be "black-and-white" images which are then pseudocolored using programs such as Adobe® Photoshop. Such images can be stored readily on compact discs, assembled into figures, and printed out with a dye-sublimation printer. Electronically stored images also can be transmitted to other sites using computer networks.

F. Troubleshooting

Poor tissue quality on sections

Brittle tissue which cracks upon sectioning usually indicates excessive time in xylene or too high temperature of the wax. If air bubbles are associated with the tissue, it is an indication that the tissue was moved too rapidly through the ethanol/xylene/wax solutions. Using old wax can also result in poor tissue embedding.

Wrinkled sections, sections coming off of the slides

Wrinkled sections frequently result when the wax block is cut too quickly or static electricity prevents the ribbon from moving away from the blade. Either slowing down the speed with which the blade passes through the knife or wetting a Kimwipe and touching it to the wax block to decrease static electricity usually helps. Sections can come off the slides during the protocol when they were wrinkled during sectioning. In addition, inadequate postfixation with PFA may be the cause; be sure to use fresh PFA.

Low transcription from template

First check that the RNA pellet is completely resuspended by vortexing and heating slightly (50°C). Second, check that the appropriate polymerase was used for a given template. Also, check whether the probe is intact by running a small aliquot out on a gel (using care to prevent RNA degradation by RNases). Lastly, extensive amounts of secondary structure, which permits intramolecular annealing in the template DNA, may prevent a good yield of high specific activity probe. In this case, a better probe may be obtained by using a shorter template or by using another region of the template.

High background

This can result from a variety of problems. A wash temperature that is too low is a frequent problem. Try increasing the temperature 5 degrees in both the hybridization and wash steps. A major source of background signal can result when using antisense RNA probes that are contaminated with a small amount of sense strand RNA. A common source of such sense strand synthesis is the initiation of the RNA polymerase from the ends of the template DNA. Such aberrant initiation is more frequent when the ends of the template have a 3′ overhang. If even a small amount of the probe is double stranded, it will be resistant to degradation by RNase A. Sense RNA can be isolated away from antisense RNA by gel electrophoresis as described in Reference 11.

When using ^{35}[S]-labeled probes, either DTT or beta-mercaptoethanol is required to prevent oxidation of sulfhydryl groups. Both agents are heat labile. Therefore, it is advisable to add these compounds just before adding the slides to a heated solution, using them once and discarding the unused portion.

High background in emulsion

Usually this results from exposure of the emulsion to light or beta radiation. Check that the appropriate safelights are being used and the box in which the slides are being dried is light-tight. Be aware of watches that glow in the dark or beepers with lights. The distance between the safelight and the emulsion should be a minimum of 4 feet.

Emulsion removal

At times, it may be necessary to remove the emulsion layer from the slide — for example, if the unprocessed emulsion layer has been exposed unintentionally to light or the processed emulsion layer has high background and the specimens are of sufficient value to warrant recovery. Although it is difficult to remove the emulsion without some damage to the underlying tissue because of the use of alkaline solutions, the following protocols will allow removal of most of the emulsion:

1. If the emulsion has not been hardened by photographic developing, it is reasonably soft and can be removed by placing the slides in a 60°C waterbath for a few minutes to soften the emulsion. Rinse the emulsion off by gentle agitation in the waterbath or under running tap water. Place the slides in Kodak fixer for 3 minutes to remove the faint bluish stain, then wash for 5 minutes in running water to remove all traces of the fixer. After being dried completely, the slide can be re-dipped in emulsion.
2. If the emulsion layer has been hardened by photographic processing, it is much more difficult to remove it. First, digest the slide for 2 to 5 minutes in a 1% solution of NaOH or KOH at room temperature. Any remaining silver grains in contact with the tissue can be removed by placing the slides in a silver reducer such as 7.5% potassium ferricyanide. This is followed by a water rinse, then 3 minutes in Kodak fixer and 5 minutes of water washing. The slides then can be re-dipped.

IV. *In Situ* Hybridization on Whole Mounts[3-5]

A. Preparation and Synthesis of Riboprobes

Digoxigenin-labeled riboprobes are prepared in a similar way as radioactive-labeled riboprobes (Section III.C).

Template and Transcription Reaction

It is better to have long inserts (≥0.6 kb) to increase the sensitivity of detection of transcripts which are expressed broadly or expressed at low levels. The procedure described here will work well with riboprobes ranging from 0.3 to 1.2 kb.

Single-stranded riboprobe is prepared by the transcription of linearized plasmid template using digoxigenin-tagged rUTP in the substrate mix. We include 0.2 μCi of α-^{32}[P]rUTP as a tracer to determine the yield of the reaction.

- 4 μl of 5× transcription buffer (5× is 200 m*M* Tris-HCl pH 8.0, 40 m*M* $MgCl_2$, 250 m*M* NaCl, 10 m*M* spermidine)
- 1.5 μl of 1 μg/μl of restricted, proteinase K-treated, linearized DNA template
- Add sufficient DEPC-treated water to make a final volume of 20 μl.
- 2 μl of 10 m*M* rATP
- 2 μl of 10 m*M* rCTP
- 2 μl of 10 m*M* rGTP

- 1.3 μl of 10 m*M* rUTP (can be replaced by dig-11-rUTP)
- 0.7 μl of 10 m*M* dig-11-rUTP
- 0.5 μl of RNAsin (40 U/μl)
- 2 μl of 0.1 *M* DTT (use DTT that has been thawed once)
- 1 μl diluted α-^{32}[P]rUTP (0.2 μCi per μl)

Mix well, spin, then add 40 U of T3 or T7 RNA polymerase. Incubate at 37°C for 2.5 to 3 hours. Digest template by adding 2 μl of DNase I and incubate for 15 minutes at 37°C. Take 1 μl aliquot of synthesized probe to measure yield of incorporation (probe can be stored on ice or at –20°C meanwhile).

Determination of Incorporation

Spot two 0.5 μl aliquots of reaction on two Whatman GF/C glass-fiber filters and allow filters to air dry. Transfer one filter into 200 to 300 ml ice-cold 5% TCA containing 20 m*M* sodium pyrophosphate. Swirl the filters in this solution for 2 minutes. Wash two more times. Transfer the filter to 70% EtOH, wash briefly, then let it air dry. This filter and the nonwashed filter are placed into scintillation vials, cocktail is added, and the radioactivity is measured in the ^{32}[P] channel. 100% incorporation of all precursors, with substrate concentrations as specified above, should give 26 μg RNA. Yields generally range from 35 to 45%, but one can use probes even if incorporation is ≈10 to 20%. A transcription reaction should synthesize about 10 μg probe.

Probe Purification

Precipitate the riboprobe by adding 10 μl 7.5 *M* NH_4OAc and 60 μl 100% ice-cold EtOH. Precipitate at –20°C, wash, air-dry the pellet as usual (see also Section III.C). Resuspend in 10 μl distilled water, then add hybridization buffer to make a solution of 10 μg/ml. For this and the subsequent steps, water should be sterilized but not DEPC treated, since this agent interferes with some of the later steps. Probe can be stored at –20°C for months.

B. Fixation of Embryos

Dissect embryos in ice-cold PBS. Remove the tissues (extraembryonic membranes) that will trap reagents. In addition, open up cavities (brain, heart) which would otherwise trap reagents. Embryos are fixed for 90 minutes in MEMFA at room temperature. Embryos are then rinsed once in 0.9% NaCl for 2 minutes and transferred to 90% MeOH, a medium in which they can be stored at –20°C. In these and the subsequent manipulations, we use 1 ml transfer pipets. Embryos become very sticky and fragile after proteinase K treatment; be careful not to damage the embryo when removing liquids or when embryos must be transferred. Use a black background to visualize the embryos.

C. Rehydration, Proteinase K Treatment, Postfixation, and Acetylation

Rehydration

Embryos are rehydrated by passage through the following solutions for 10 minutes each step:

1. 75% methanol/H_2O
2. 50% methanol/H_2O
3. 25% methanol/75% PBT (PBT = PBS, 0.1% Tween 20; first dilute detergent to 10% with water)
4. PBT twice

Proteinase K Treatment and Postfixation

Transfer embryos to a small petri dish (3.5-cm diameter) containing PBT at room temperature. Remove most liquid and add freshly prepared protease K solution (10 μg/ml) to the dish. Embryos should be well immersed and incubated for 3 to 20 minutes according to their dimension (e.g., chick embryos, Hamburger-Hamilton (HH) stage 4, 3 minutes; HH stage 8, 6 minutes; HH stage 22, 20 minutes). Protease batches can vary in strength; thus, incubation times should be optimized empirically. The protease stock should be aliquoted and stored at –20°C. Following proteinase K treatment, rinse twice with PBT. Transfer into 5 ml 100 m*M* glycine (in PBT; adjust pH to 7.2) and incubate for 5 minutes at room temperature. Wash twice in PBT, 5 minutes each. Embryos are then refixed for 30 minutes at room temperature in 4% paraformaldehyde-PBS containing 0.1% Tween (0.2% glutaraldehyde may be added to this fixation if the tissue is particularly fragile, such as early embryos). This is followed by two 5-minute washes in 5 ml of PBT.

D. Prehybridization and Hybridization

Transfer the embryos into 7-ml glass vials with Teflon-lined caps. Remove most of the PBT solution, then underlay the residual PBT with 2 ml of hybridization buffer. Allow embryos to sink into the hybridization solution (may take 20 to 30 minutes), remove as much of the upper layer (PBT) as possible, then mix. After 10 minutes, remove the hybridization buffer, then add 2 ml fresh hybridization buffer and incubate for 4 hours at 60°C in a very gently rotating water bath (18 rpm). Place vials at room temperature for 5 minutes and carefully remove the hybridization solution. Add 1.5 ml hybridization buffer containing 0.2 μg/ml digoxigenin riboprobe (1/50 volume of stock). Thoroughly mix by swirling (or inverting if a Teflon-lined cap is used) and incubate the embryos overnight in a 60°C water bath. Smaller volumes of hybridization buffer can be used and the probe mix can be reused once.

Recipe

Hybridization buffer

- 50% deionized formamide
- 5× SSC
- 5 m*M* EDTA
- 1 mg/ml yeast tRNA (purified with phenol/chloroform, followed by EtOH precipitation)
- 100 μg/ml heparin, 1× Denhardt's solution
- 0.1% Tween 20
- 0.1% CHAPS (3-[3-cholamidopropyl)-dimethyl-ammonio]-1-propanesulfonate, Boehringer)

E. Posthybridization Washes, RNase Treatment, and Post-RNase Washes

Posthybridization Washes

Unless otherwise specified, the washes are performed in a 60°C water bath rotating at a rate of 18 rpm. The washes should be thorough but gentle. It is important to allow embryos to settle to the bottom of the tube and to leave some liquid above them when removing the wash solutions; otherwise, the embryos will be flattened and damaged.

1. Rinse embryos twice in 5 ml FSC (50% formamide; 2× SSC; 0.1% CHAPS; 50 m*M* glycine, pH 7.2) at room temperature.
2. Wash twice in 5 ml FSC, 30 minutes per wash.
3. Wash twice in 5 ml 2× SSC-CHAPS (2× SSC; 0.3% CHAPS; 50 m*M* glycine, pH 7.2), 5 minutes per wash.
4. Resuspend in 1.8 ml SSC-CHAPS (1× SSC; 0.3% CHAPS; 50 m*M* glycine, pH 7.2) at room temperature.

RNase Treatment

The reaction is carried out for 35 minutes at 37°C in SSC-CHAPS containing 4 μg/ml RNase A (RNase A is dissolved in water at a concentration of 1 mg/ml and frozen at –20°C in aliquots; do not boil the stock) and 20 U per ml RNase T1. Specifically, mix in an Eppendorf tube 191.6 μl SSC-CHAPS, 0.4 ml RNase T1 (100,000 U/ml), and 8 μl RNase A (1 mg/ml) and add this RNase cocktail to the embryos. Mix gently and incubate 35 minutes at 37°C, agitating occasionally.

Post-RNase Washes

1. Wash once in 5 ml 2× SSC-CHAPS at room temperature for 5 minutes.
2. Wash briefly in 5 ml FSC at room temperature.

3. Wash twice in 5 ml FSC at 60°C with gentle agitation, 30 minutes per wash.
4. Wash once in 5 ml 2× SSC-CHAPS at 60°C for 5 minutes.
5. Wash twice in 0.2× SSC-CHAPS (0.2× SSC; 0.3% CHAPS) at 60°C for 15 minutes.
6. Wash once in PBT-CHAPS (PBS; 0.1% Tween 20; 0.3% CHAPS) at 60°C for 5 minutes; cool to room temperature.

Note: 60°C works well for most of the probes. If the signal is rather strong and localized, RNase treatment and post-RNase washes can be substituted by two washes with 0.2× SSC-CHAPS for 30 minutes at 55 to 65°C and then a single wash with PBT-CHAPS for 5 minutes at 55 to 65°C. However, if this shortcut is applied, the temperature for hybridization has to be optimized.

F. Antibody Incubation and Washes

To visualize the digoxygenin hapten, a high-affinity antibody from Boehringer is used. To prevent nonspecific binding of the antibody, embryos are pre-blocked with lamb serum. In addition, the polyclonal antibody may be pre-absorbed to embryo powder or to embryos, but this step is not always required.

1. Wash three times in 5 ml PBTB (PBS; 0.1% Triton X-100; 2 mg/ml BSA).
2. Add 1 ml of PBTB containing 40% lamb serum which has been heat inactivated at 55°C for 30 minutes. Block for 1 hour at 4°C, gently rocking the specimens.
3. Replace liquid with 2 ml of PBTB-40% lamb serum containing a 1/2000 dilution of sheep anti-digoxigenin Fab antibody conjugated to alkaline phosphatase. This solution can be reused more than once. Include 0.1% sodium azide to prevent bacterial growth during storage at 4°C.
4. Incubate overnight at 4°C with gentle rocking.
5. Rinse briefly with PBTB 3 times at room temperature.
6. Rinse four times with PBTB for 1 hour per wash, with gentle agitation on a rocking platform.

G. Alkaline Phosphatase Color Reaction

1. Rinse the embryos briefly with 4 ml alkaline phosphatase buffer twice (100 m*M* Tris, pH 9.5; 100 m*M* NaCl; 50 m*M* $MgCl_2$; 0.1% Tween 20).
2. Add 3 ml alkaline phosphatase buffer into the vial and wrap it with aluminum foil to prevent light exposure. For color reagents, mix 1 ml alkaline phosphatase buffer with 18 µl nitro blue tetrazolium (75 mg/ml in 75% dimethylformamide) and 14 µl BCIP (5-bromo-4-chloro-3-indolylphosphate, 50 mg/ml in dimethyl-formamide). Add this color reagent mixture to the embryos and gently agitate in the dark for 2 to 9 hours at room temperature.
3. Observe the development of the stain briefly every half hour and try to catch the best staining (strong signal and low background). Slightly overstain the embryos, as some of the background stain will be washed off in the subsequent

treatments. The embryos should be examined only briefly, since prolonged illumination will turn the yellow reaction mixture brown which may increase the background.

4. Stop the color reaction by postfixation of the embryos in MEMFA for 30 minutes at room temperature, then rinse specimens in 0.9% NaCl for 2 minutes and transfer to 90% MeOH. If the background is high, subject embryos to several washes in 90% MeOH at 4°C with gentle rocking (overnight to 2 to 3 days). The stained embryos can be stored at –20°C. Embryos can be restained if they have not been postfixed by MEMFA. The embryos that require restaining are kept in PBT at 4°C.

H. Embedding and Sectioning Whole-Mounted Embryos

The stained embryos can be photographed in a dissecting microscope with bright- or darkfield illumination. They subsequently can be embedded and sectioned to examine the details of the expression patterns. Although the reaction product using NBT/BCIP is soluble in many organic solvents, it is stabilized by fixation with formaldehyde, and the embryos then can be embedded in paraffin wax or Histo Prep (Fisher Scientific) in the case of frozen sections.

1. Paraffin Sections

1. Transfer the embryos to 100% EtOH for 10 minutes at room temperature.
2. Transfer to xylene for 5 minutes, twice.
3. Transfer to xylene/paraffin wax (prewarmed at 58°C) for 1 hour at 58°C.
4. Transfer to paraffin wax (prewarmed at 58°C) and incubate overnight at 58°C.
5. Embed the embryo as described in Section III.A.
6. The embryo is sectioned at 10-μm thickness and the sections are placed on Superfrost slides (Fisher) as described in Section III.B.
7. Dewax the slides with xylene for 10 minutes, twice, and then progressively rehydrate until sections are in water.
8. Mount the slide with Biomeda's Crystal/Mount (cat. no. M02). Place the slides horizontally in an oven set at 70 to 80°C for at least 10 minutes. Remove the slides and allow them to reach room temperature. These slides are now ready for microscopy.

2. Frozen Sections

1. Rehydrate the embryos progressively to PBS and rinse with PBS briefly three times.
2. Let the embryos lie on top of 30% sucrose/phosphate buffer (10 m*M* Na_2HPO_4, 1.8 m*M* KH_2PO_4, pH 7.4). Keep at 4°C until the embryos sink to the bottom.
3. Put the embedding mold on top of smashed dry ice and put some OCT into this cooled mold. Place and orient the embryo in the OCT, and wait for it to solidify.

4. When OCT is hardened, cut away the extra OCT. Put some OCT on top of the sample holder of the cryostat at room temperature and place the trimmed tissue block on top of it. The tissue block should be glued by the OCT. Put the sample holder rapidly into crushed dry ice and let it cool so that the tissue block firmly attaches to the holder.
5. Section at 10-μm thickness and mount these sections on Superfrost slides (Fisher) as described in Section III.B.
6. Air dry the slides completely.
7. Cover the entire slides with PBS for 1 minute and remove PBS by touching the margins on top with paper towels.
8. Put three drops of the Biomeda Crystal/Mount (cat. no. M02) and mount the slide as described above.

Acknowledgments

We would like to thank Drs. Ariel Ruiz i Altaba, Clementine Hofmann, Beat Lutz, Olof Sundin, and Mr. Calvin Wong, who have contributed to the development of these protocols. This work was supported in part by funds from the National Institutes of Health (GE, JH) and from the Swiss National Science Foundation (UA).

References

1. Koopman, P., Analysis of gene expression by reverse transcriptase-polymerase chain reaction, in *Essential Developmental Biology. A Practical Approach,* Stern, C.D. and Holland, P.W., Eds., Oxford University Press, Oxford, 1993, 233–242.
2. Holland, P.W., Cloning genes using the polymerase chain reaction, in *Essential Developmental Biology. A Practical Approach,* Stern, C.D. and Holland, P.W., Eds., Oxford University Press, Oxford, 1993, 243–256.
3. Wilkinson, D.G., *In situ* hybridization, in *Essential Developmental Biology. A Practical Approach,* Stern, C.D. and Holland, P.W., Eds., Oxford University Press, Oxford, 1993, 257–276.
4. Wilkinson, D.G., *In Situ Hybridization. A Practical Approach.* Wilkinson, D.G., Ed., Oxford University Press, Oxford, 1992.
5. Li, H.-S., Yang, J.-M., Jacobson, R.D., Pasko, D., and Sundin, O., *Pax-6* is first expressed in a region of ectoderm anterior to the early neural plate: implications for stepwise determination of the lens, *Dev. Biol.,* 162, 181–194, 1994.
6. Sambrook, J., Fritsch, E.F., and Maniatis, T., *Molecular Cloning. A Laboratory Manual,* Cold Spring Harbor Laboratory Press, New York, 1989.
7. Reiner, O., Albrecht, U., Gordon, M., Chianese, K.A., Wong, C., Gal-Gerber, O., Sapir, T., Siracusa, L. D., Buchberg, A. M., Caskey C. T., and Eichele, G., Lissencephaly gene (*LIS 1*) expression in the CNS suggests a role in neuronal migration, *J. Neurosci.,* 15, 3730–3738, 1995.

I. Introduction

The development of a multicellular organism from a fertilized one-cell embryo is orchestrated by complex cellular, molecular, and genetic processes. Genetic engineering strategies have advanced our understanding of these processes by facilitating the cloning, sequencing, and cellular localization of numerous gene products, such as transcription factors and growth factors. The transfer of genes into cultured cells has provided useful information concerning the necessary machinery for gene transcription in certain cell types and has provided insight into the function of various gene products in these cells. However, it is not possible to study many developmental processes *in vitro*. In the past decade, gene-directed manipulations in genetically engineered mice have provided new avenues for gathering *in vivo* information concerning the regulation and function of genes during mammalian development.

An organism into which a gene has been transferred is referred to as a transgenic organism.[1] An important feature of transgenic organisms is that they enable the investigation of gene regulation and function in the context of the whole animal and during the normal developmental program.[2] The importance of cell to cell interactions and the interaction of cells with their environment make this feature paramount. The study of gene regulation and function in complex biological systems, such as a developing embryo, have benefited greatly from the advancement of transgenic technology and the development of transgenic mice. The success of gene-directed manipulation in mice depends on exogenous DNA being introduced into the germline of mice. This is accomplished either through zygote microinjection[1-4] or by injection of genetically modified embryonic stem (ES) cells into blastocysts.[3-5] The latter is one of the most important scientific advances in recent years in that it allows for the introduction of specific gene modifications into ES cells, which can go on to be incorporated into the germline of mice.[6-9] This manipulation of the germline through the genetic modification of ES cells is a valuable *in vivo* means of investigating aspects of mammalian development, protein function, and disease processes.

Transgenic mice should be considered as an important resource for all scientists, regardless of prior experience with molecular techniques. The use of transgenic technology is no longer limited to laboratories with the resources and technology to generate transgenic mice. Most institutions now contain facilities for the generation of transgenic mice, or the generation of mice can be contracted through commercial organizations or through collaborative efforts. In addition, many mice that have already been generated are available commercially or through collaborative means. The limiting factor for researchers is no longer the generation of transgenic mice, but rather the generation of ideas and questions which can be addressed by this powerful technology. Excellent detailed protocols for generating transgenic mice are available,[3-5] so protocols will not be discussed here. Rather, this chapter is designed to familiarize the reader with the general methodology used in creating transgenic mice and to provide examples of how they have been utilized to study mechanisms of normal and abnormal development, tumorigenesis, and disease.

II. Zygote Microinjection

The expression of exogenous genetic information in mice is most commonly accomplished through zygotic microinjection.[1,2] In this technique, an exogenous gene is microinjected into the pronuclei of one-cell fertilized embryos (zygotes) which are subsequently implanted into pseudopregnant foster mothers (Figure 3.1).[3] Typically, several to many copies of the exogenous DNA will randomly integrate into the genome in a nonhomologous manner at a single site. With subsequent cell division, the DNA is distributed to all cells of the developing embryo including the germ cells, thus producing founder mice which are hemizygous for the transgenic locus. Founder mice are identified by obtaining genomic DNA by tail biopsy followed by detection of the transgenic locus using DNA hybridization (Southern blot or dot blot analysis) or polymerase chain reaction (PCR). Most transgenes will contain DNA sequences that are either foreign to mice (i.e., bacterial sequences found in commonly used reporter genes) or are arranged in a manner dissimilar to the endogenous gene (this is the case with many minigenes). These features aid in the identification of the transgenic locus and in the determination of the number of transgenes integrated (copy number). Founder mice subsequently are bred to establish lines that can be used to analyze expression from the transgenic locus and/or assessment of phenotypes associated with transgene expression. If it is necessary to identify transgenic embryos, genomic DNA can be isolated from extraembryonic membranes of individual embryos or fetuses.

This procedure allows for the introduction of exogenous DNA into the mouse genome. The investigator can use this technology to address specific questions depending on the nature of the DNA fragment integrated. Many genes are expressed in specific cell types at varying levels and at different stages of development. Transgenic mice can be used to identify the genetic regulatory elements necessary to establish these specific patterns of gene expression.[2] Once identified, these elements can be used to target the expression of other genes to specific tissues to assess the function of that gene product in specific cells during development. This is a powerful way to study the regulation and function of genes during development, and specific examples will be discussed below. There are, however, some important caveats which must be considered when using transgenic mice generated by zygote microinjection. The integration of DNA into the genome following zygote microinjection is random. This is a potential problem in that transcription from the transgenic locus can be affected by where it integrates into the genome. The DNA may be integrated into a region of the genome which, due to chromatin makeup, is less susceptible to transcription and may influence the transcription

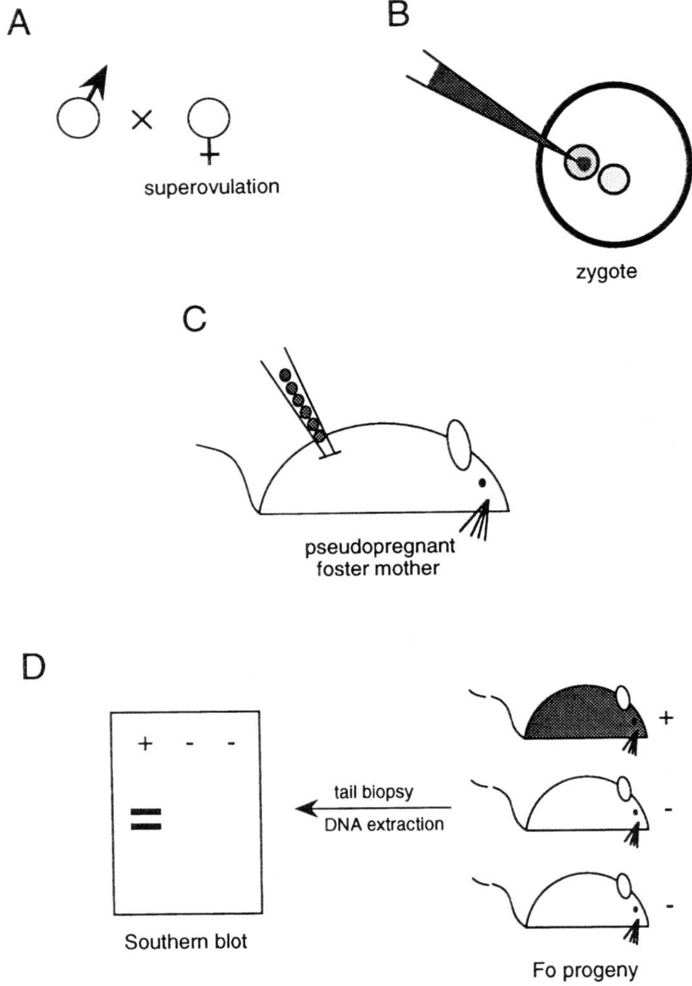

FIGURE 3.1
Zygote microinjection. (**A**) Female mice are treated with hormones to stimulate superovulation and are then mated with fertile males. Zygotes are collected from oviducts of pregnant females on 0.5 days postcoitum and treated with enzymes to remove cumulus cells. (**B**) Using a microscope equipped with micromanipulators, the zygote is held in place by a holding pipette while DNA suspended at a concentration of 2 to 5 ng/μl is injected into the male pronucleus. The injected transgene usually will integrate into the genome before the two-cell stage, ensuring the inheritance of the transgene to all cells. (**C**) Zygotes which survive microinjection are placed surgically into the oviducts (usually 12 zygotes per oviduct) of foster mothers rendered pseudopregnant by mating with vasectomized males. (**D**) Transgenic Fo progeny are identified at weaning (approximately 4 weeks of age) by extracting genomic DNA from a small piece of tail and screening for the presence of the transgene using Southern blot, dot blot, or PCR. These assays also will provide information concerning the number of transgenes integrated into the genome. When screening for transgenic embryos or fetuses, genomic DNA can be isolated from extraembryonic membranes. Positively identified transgenic mice can then be bred to establish lines to study the expression pattern of the transgene or associated phenotypes. Detailed procedures for the generation of transgenic animals by zygote microinjection are available.[3,4]

efficiency from the transgenic locus. Likewise, transcription from the transgenic locus may be influenced by regulatory elements of endogenous genes near the site of integration. These effects are referred to as "position effects." These problems can be overcome by generating several independent transgenic lines. Another potential problem associated with the random nature of DNA integration during zygote microinjection is integration of the transgenic locus within an essential gene. Such an insertion may result in lethality when the transgenic animals are bred to homozygosity. These types of insertional disruptions occur in less than 10% of transgenics and can be overcome by generating multiple independent lines. These caveats are minor relative to the information to be gained through the successful transfer and analysis of genes in the whole animal.

A. Gene Regulation

The mammalian genome consists of approximately three billion base-pairs of DNA sequence information organized into an estimated 100,000 genes. Available evidence suggests that approximately 25,000 genes are expressed in all tissues and that the remaining 75,000 genes are expressed in a tissue restricted manner. The transfer of genes into cells *in vitro* has been useful in identifying many features of cell-type-specific expression of genes; however, such studies are limited in their ability to assess the genetic mechanisms required for developmental gene expression. In addition, it is often found that cells in culture lack the environmental and cellular signals necessary to recapitulate normal patterns of gene expression. When using transgenic mice, the transgene is present in every cell of the embryo and adult, allowing the study of gene expression in all cells throughout the developmental program.[2,3] These features make transgenic mice an excellent tool for deciphering gene regulatory elements necessary for tissue specific and developmental gene expression.

If a gene is highly expressed in a certain cell type during development, the genetic mechanisms responsible for this expression can be deciphered by testing the ability of DNA that is flanking or within this gene to direct expression of a reporter gene in transgenic mice (Figure 3.2). A reporter gene is any gene for which expression can be monitored. A commonly used reporter gene is the bacterial *lacZ* gene encoding β–galactosidase, which is easily monitored histochemically. Other reporter genes include chloramphenicol acetyltransferase (CAT) and luciferase. Typically, the promoter and various sized DNA fragments from the 5′ flanking region of the gene of interest are fused to the reporter gene, which is then microinjected into zygotes. Transgenic mice carrying this construct are identified and the pattern of reporter gene expression monitored (Figure 3.2). If the proper developmental expression pattern for this gene is captured, then subsequent DNA constructs containing deletions of the larger original fragment can be tested in a similar manner, thus defining the minimal *cis* regulatory elements necessary for the cell-type-specific and developmental expression of this gene. This type of genetic analysis has led to the

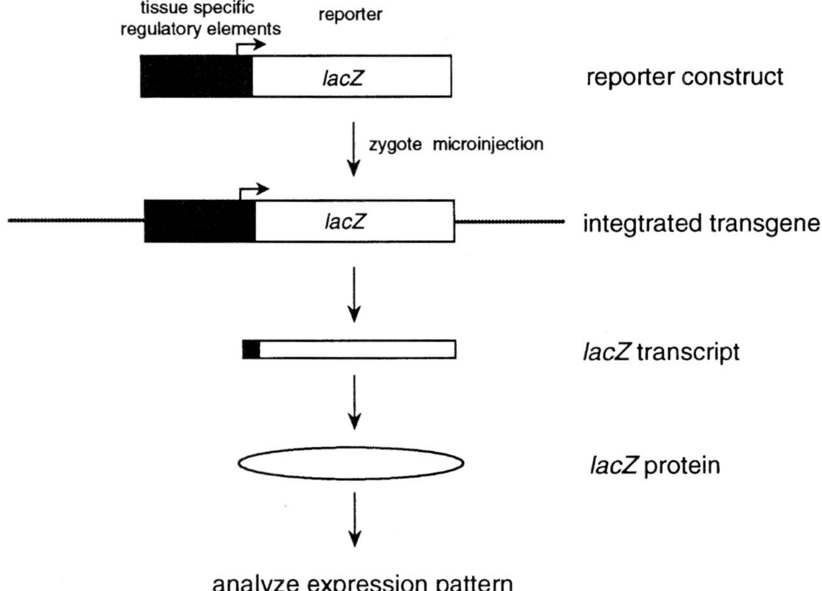

FIGURE 3.2

Tissue-specific targeting of transgenes. DNA regulatory elements and a cDNA are fused together in a prokaryotic cloning vector. The regulatory elements used to drive expression of the transgene include a tissue specific promoter and typically additional regulatory sequences necessary for proper tissue specific and developmental regulation. Depicted here is a fusion construct harboring the bacterial *lacZ* gene. *LacZ* is commonly used as a reporter because β-galactosidase is easily visualized by an enzymatic assay. Similar fusion constructs can be made using other DNA regulatory elements and cDNAs, depending on the experimental intent. The DNA construct is purified away from the cloning vector and microinjected into zygotes, where it will integrate into the genome in a nonhomologous manner. In response to the proper developmental and cellular cues, the integrated transgene will be transcribed into messenger RNA (the arrow denotes the start and direction of transcription) and subsequently functional protein. In this manner the tissue-specific expression of reporter genes or functional genes can be analyzed in the context of the whole animal.

identification of hundreds of gene regulatory elements responsible for expression in many tissues throughout the developing embryo and adult. Identification of *cis* regulatory elements is often a critical first step towards the identification of *trans* acting factors which are necessary for tissue specific and developmental gene regulation. In addition to facilitating the characterization of *cis* regulatory elements and *trans* acting factors, transgenic mice have allowed for the identification of higher order regulatory domains which appear to influence chromatin configurations around genes and to facilitate their expression. These regions include locus control regions[10] and matrix attachment regions[11] which were unmasked by the ability to manipulate gene expression in transgenic mice.

B. Ectopic Expression

Once promoters and gene regulatory elements necessary for tissue-specific and developmental gene regulation have been identified, they can be used to address a number of biological questions in transgenic mice. The diverse tissue-specific regulatory elements that have been identified allow for the targeting of any gene of choice to nearly any cell or tissue in the embryo or adult. DNA constructs consisting of tissue-specific promoters and regulatory sequences are fused to cDNAs (Figure 3.2). These constructs are then microinjected into zygotes to generate transgenic mice expressing exogenous cDNAs. This genetic manipulation in mice provides an *in vivo* means of altering what tissues and cells are doing with regard to what proteins are being produced. These types of manipulations fall into either a gain of function or loss of function category, of which there are many examples.

1. Gain-of-Function Transgenic Mice

An example of gain-of-function approaches is seen in studies in which the expression of turmorigenic factors, such as the simian virus 40 small tumor antigen or oncogenes such as *myc* and *ras*, are directed to various cells in order to transform them *in vivo*.[12,13] Such studies have provided models for investigating the mechanisms of tumorigenesis in various cell types and have provided a means of obtaining permanent cell lines from these oncogenic cells.[14,15] Ectopic genes or the overexpression of genes in transgenic mice have provided useful animal models for studying and treating various diseases. Erythroid-specific DNA elements were utilized to overexpress the human β^s-globin gene in transgenic mice.[16,17] Doing so produced mice which exhibited features of sickle cell anemia in humans. A neural specific enolase promoter was used to target the expression of β-amyloid precursor protein in transgenic mice to create animals with β-amyloid protein deposits in their brains,[18] a feature common to Alzheimer's disease. Transgenic technology also can be useful for overproducing and purifying proteins. An ideal organ for such production is the mammary gland, where, under the control of β-lactoglobulin DNA regulatory elements, large amounts of the protein of choice can be made and collected from the milk without affecting the donor. This has been demonstrated successfully for the over-production of human α_1-antitrypsin, a protein used in the treatment of emphysema, in transgenic mice and sheep.[19-21] This type of biopharmaceutical technology has profound implications for the medical and livestock industries. Interestingly, transgenic mice have been engineered to produce human antibodies against human target antigens,[22] which has significant clinical implication in the mass production of human antibodies. There are many other examples of gain-of-function uses for transgenic mice, including the tissue-specific ectopic or overexpression of regulatory molecules such as peptide hormones or receptors, transcription factors, or other components of signaling pathways. These strategies clearly have provided new insight into the physiological role of many regulatory factors and signaling pathways, and the mice already generated should be considered as a resource to investigators.

2. Loss-of-Function Transgenic Mice

Transgenic mice generated by zygote microinjection also can be utilized in the formation of loss-of-function mutations in which molecules designed to inhibit the normal function of an endogenous gene are targeted to specific cell types using specific DNA regulatory elements. One example of this type of strategy is the expression of antisense RNA specific for the target gene. Antisense RNA, in turn, will form duplexes with the endogenous sense RNA causing its destruction and therefore preventing translation. This strategy was demonstrated successfully in a study where virus-induced leukemia was inhibited in transgenic mice in which lymphocyte DNA regulatory elements were used to target expression of antisense RNA complementary to the retroviral packaging sequences of the Moloney leukemia virus.[23] Other loss-of-function mutations have focused on dominant-negative approaches, in which the expression of defective receptors can function to suppress the activity of wild-type receptors. Epidermis-specific DNA regulatory elements isolated from the human keratin 1 gene were used to target expression of dominant-negative retinoic acid receptor to the skin of transgenic mice.[24] Doing so resulted in neonatal lethality due to the loss of skin barrier function. In another study, a dominant-negative mutant of the human insulin receptor was overexpressed in muscle of transgenic mice under the control muscle-specific DNA regulatory elements from muscle creatine kinase.[25] This resulted in impaired insulin responsiveness in these mice which can serve as models for tissue-specific insulin resistance.

Other loss of function mutations can be generated to address the function or importance of a particular cell type. Specific cells types can be eliminated by expressing toxic molecules in transgenic mice using tissue-specific DNA regulatory elements. A commonly used toxin is diphtheria toxin, a bacterial gene which encodes a protein which inhibits translation.[26] Targeted expression of the diphtheria toxin gene in transgenic mice using regulatory elements from the γ_2-crystalline gene resulted in the ablation of lens fibers and the generation of such eye disorders as cataracts, microphthalmia, and anophthalmia.[27,28] Other cell types subjected to this type of analysis include pancreatic acinar cells using elastase DNA regulatory elements,[29] and somatotrope and lactotrope cells of the anterior pituitary gland were selectively destroyed by expressing diphtheria toxin under the control of DNA regulatory elements from the rat and human growth hormone gene.[30] Interestingly, ablation of these cells resulted in decreased production of growth hormone and the generation of dwarf mice. In these examples, cells expressing diphtheria toxin were killed whenever the gene regulatory elements being used were activated. An alternative approach is to target the expression of a gene for which toxicity is inducible. This has been accomplished using the herpes simplex virus 1 thymidine kinase (*HSV-1-tk*).[31] The expression of *HSV-1-tk* itself is not toxic to mammalian cells; however, it is capable of metabolizing the otherwise non-toxic nucleoside analog ganciclovir, which in turn leads to inhibition of DNA synthesis and cell death. Thus, it is possible to express *HSV-1-tk* in a tissue-specific manner in transgenic mice and then induce the destruction of these cells by treating the animals with ganciclovir. This strategy has been used to induce the ablation of T and B cells expressing *HSV-1-tk* under the control of lymphoid specific DNA regulatory elements.[32a] This conditional ablation of cell types is not only useful in assessing the physiological importance of cell populations, but it also provides a means of studying the interrelations of cell lineages during growth and development.

Studying the physiological importance of a gene or cell population during development or disease process would benefit from the ability to control precisely when a transgene is expressed. Recently. a highly efficient steroid-inducible system has been described, in which transgenes under the control of an ecdysone responsive promoter can be induced upon administration of the hormone.[32b] Ecdysone is an insect hormone not found in mammals and therefore has no side effects. This system will be useful in controlling the induction of both gain of function and loss of function transgenes.

III. Embryonic Stem Cells

Perhaps the most significant advancement in transgenic technology has been the advent of embryonic stem (ES) cells, and the demonstration that they can contribute to the germ line of chimeras.[6-8] ES cells are derived from totipotent inner cell mass cells of blastocysts. During normal development these cells form all the cell types found in the embryo proper. Under specific conditions these cells can be maintained in culture, where they proliferate but remain diploid and do not lose their potential to differentiate into embryonic cells.[33] This feature is critical, because it has allowed for sophisticated genetic manipulation of these cells in culture; moreover, it has also allowed for the selection of ES cells which have undergone the desired genetic changes (Figure 3.3).[3-5] The genetically manipulated cells then can be transplanted into the blastocoel of blastocysts, which are subsequently transferred to pseudopregnant recipients where they go on to contribute to tissues of the developing embryo leading to the production of chimeric mice. Genetic markers such as coat color are used to aid in the analysis of chimeras. For example, ES cells derived from mice with black coat color are transplanted into blastocysts from mice with white coat color. The extent of ES cell contribution to tissues of chimeric mice can be estimated by the degree of chimeric coat color. In the event that ES cells contributed to the germ line, subsequent breeding of chimeras will lead to germ line transmission of the genetic mutation and the generation of mice that are heterozygous for the genetic mutation. In this manner it is possible to generate transgenic mice that have specific genetic alterations which can be used to assess the necessity and function of specific genes.[34]

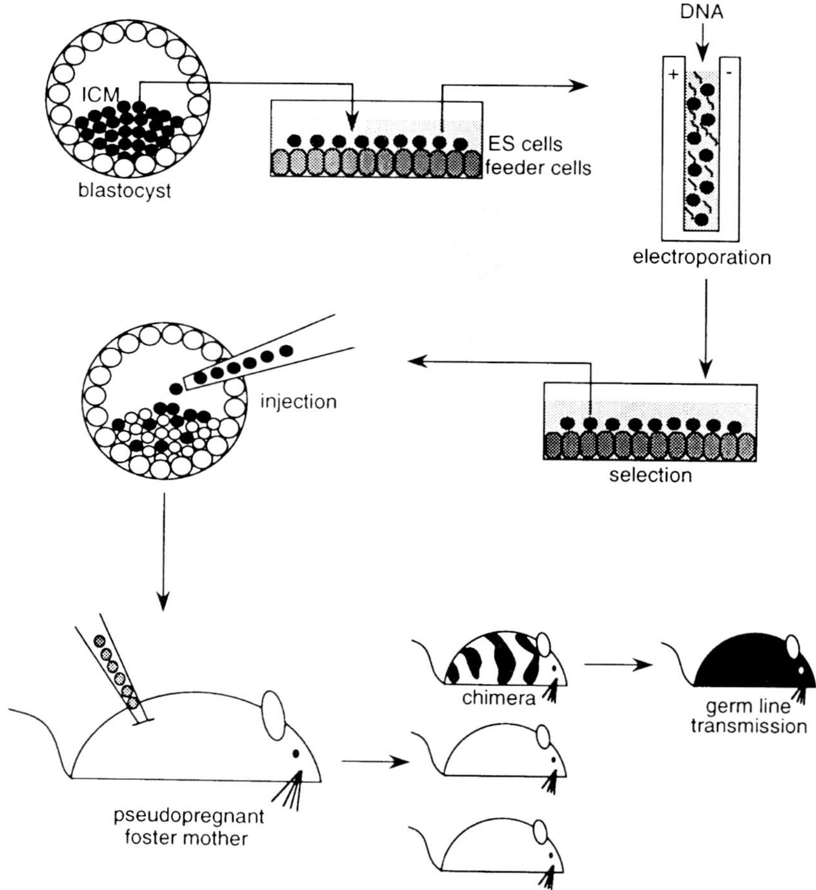

FIGURE 3.3

Embryonic stem (ES) cells and gene targeting. Blastocysts are collected from pregnant uteri on 3.5 days postcoitum and the inner cell mass (ICM) is removed and the cells placed in culture. ES cells are cultured with feeder cells which have been irradiated to prevent further cell division. ES cells are maintained at a high density and passaged every 2 to 3 days to prevent their differentiation. To introduce DNA into the ES cells, they are suspended in a electroporation cuvette together with DNA and an electric current is applied to facilitate the entry of DNA into the nucleus. The DNA introduced into the cells contains genes such as *neo* and *tk*, which will allow for the selection of ES cells which have undergone homologous recombination (see Figure 3.4). Single cell suspensions are then made from surviving colonies which have been identified to contain the designed targeted mutation. These cells are then injected into blastocysts from mice with a coat color distinct from that of the mice from which the ES cells were derived. These chimeric blastocysts then are placed into the uterine horns of pseudopregnant foster mothers. The resulting chimeric progeny are bred with the hope of germ line transmission. With germ line transmission, mice heterozygous for the targeted allele are identified by Southern blot analysis or PCR and then intercrossed to obtain mice homozygous for the targeted mutation.

A. Site-Specific Mutations

As mentioned earlier, zygotic microinjection of DNA results in random integration of DNA into the genome by nonhomologous recombination. Although this provides many opportunities to study the expression of ectopically expressed genes, it does not allow for the specific manipulation of endogenous genes. Gene targeting using homologous recombination provides a means of making predetermined site-specific mutations in ES cells, which subsequently can be used to transmit mutations to the germ line of transgenic mice. This is the method of choice for creating null mutations in genes to assess their importance and to study their physiological function in transgenic mice.[34]

Successfully targeting a specific mutation in ES cells is based on the ability of sequences within the target gene to recognize and recombine with homologous sequences on a targeting vector. The most common type of targeting vector used for the generation of null mutations in transgenic mice is the replacement type vector (Figure 3.4). These vectors contain features for specific crossover events necessary for recombination, as well as features that will allow for the selection of targeted events in ES cells. Cloned sequences homologous to the targeted gene, normally protein-encoding exons, are interrupted by the insertion of a selectable marker. A commonly used selectable marker is the bacterial neomycin phosphotransferase (*neo*) gene, which, when expressed, renders mammalian cells resistant to an amino-glycoside analog (G418). The targeting vector is introduced into ES cells by electroporation where homologous recombination events replace a segment of the endogenous gene with the introduced mutant sequences. This insertional mutagenesis is designed such that the disruption in the now targeted gene will prevent the production of functional gene product. The frequency with which these recombination events occur is low; however, millions of ES cells can be electroporated and cultured in the presence of G418, enriching for integration events. Since the targeting vector can also integrate into the genome in a random nonhomologous manner, it is necessary to screen for targeted recombination. This is done by either PCR or Southern blot analysis. Homologous recombination events can be enriched by using positive-negative selection schemes (Figure 3.4). In addition to the positive selectable marker inserted between the regions of homology (i.e., *neo*), a second negative selectable marker, such as the herpes simplex virus thymidine kinase (*tk*) gene, is attached outside the regions of homology. Expression of the *tk* gene renders cells sensitive to ganciclovir. Because the *tk* gene lies outside the region of homology, it will be lost during targeted recombination; therefore, electroporated ES cells can be cultured in medium containing G418 for positive *neo* selection and ganciclovir for negative *tk* selection. This greatly increases the probability of obtaining alleles with targeted gene disruptions. It is worth reiterating that such genetic manipulations can only be performed *in vitro*, making the availability of ES cell technology critical to the development of targeted mutations in mice.

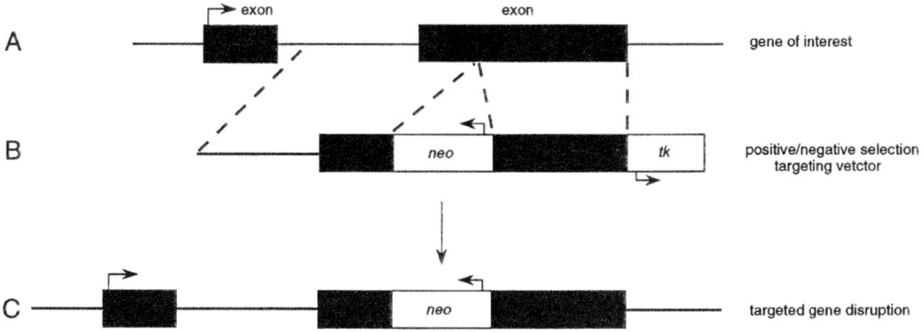

FIGURE 3.4
Gene targeting using a positive-negative selection scheme. (**A**) Depicted is a portion of the endogenous gene to be targeted by homologous recombination in ES cells. The black boxes denote exons while lines represent introns. The arrow indicates the site of transcription initiation. (**B**) Targeting vector used for positive/negative selection. The segment of the gene to be targeted is subcloned into a cloning vector and a copy of the neo gene is inserted within an exon. A copy of the *tk* gene is placed at the end. The targeting construct is then electroporated into ES cells where homologous recombination occurs and disrupts the target gene by the insertion of the *neo* gene (**C**). Because the *tk* gene is at the end of the region of homology, it will not be incorporated during homologous recombination. These events are selected for by culturing the electroporated ES cells in medium containing G418 for positive selection of cells expressing *neo* and ganciclovir for negative selection of recombination events excluding *tk* expression.

Once ES cell colonies containing the desired genetic mutation have been identified, they are amplified and microinjected into the blastocoel of host blastocysts for the generation of chimeric mice (Figure 3.4). With the incorporation of ES cells into the germ line, the targeted gene disruption is passed on to future generations. Eventually the breeding of this null allele to homozygosity will assess the biological importance of the disrupted gene within the whole animal in the context of the normal developmental program.

1. Consequences of Null Mutations

The creation of site-specific null mutations allows one to observe phenotypic consequences of selected gene disruption in heterozygous and homozygous animals. There are now hundreds of examples of genes which have been analyzed in this manner;[34] the information gained is immense and cannot be reviewed here. However, the importance of various genes for all aspects of development has been investigated by gene targeting, and information has been obtained regarding nearly all the fundamental processes of development. Often the ablation of functional gene product by targeted mutagenesis leads to a predictable phenotype, such as embryolethality or specific abnormalities consistent with the proposed function of the gene. However, this is not always the case. There are many examples of gene disruptions which result in unexpected phenotypes or no phenotype at all. Many of the genes investigated encode regulatory molecules which are part of larger families of genes which function in complex signaling pathways. Unexpected phenotypes, or the lack thereof, suggest that the function of these gene products are compensated by other molecules, thus implicating the need to investigate more closely the mechanisms involved in that particular biological process.

Mice generated by gene targeting techniques can, in turn, be used as specific genetic backgrounds to address various aspects of biology and disease. Targeted disruption of genes known to be involved in various disease processes has led to the development of animal models for human diseases. These animals will be useful in studying mechanisms of diseases as well as aid in the development of strategies for disease treatment and prevention. Depending on the questions being asked, many of the genetic backgrounds can be utilized in conjunction with pharmacological approaches to study aspects of abnormal development. An example is seen in a recent study using mice deficient in the p53 tumor suppresser gene which plays critical roles in DNA repair and apoptosis. It was shown that mice homozygous for a targeted disruption in the p53 allele were less susceptible to teratogenesis induced by DNA-damaging agents, suggesting p53 might have a fundamental embryoprotective role in DNA damage-related teratogenesis.[35b] Other mice which may be useful for studying the relative effects of teratogens on DNA damage and repair include mice with targeted mutations in NAD:protein (ADP-ribosyl) transferase (ADPRT), also known as PARP,[36] and the excision repair cross complementing gene (*ERCC-1*).[37] In another study, disruption of the multidrug resistance (*mdr1a*) gene led to a deficiency in blood brain barrier function, and the mice were found to have a increased sensitivity to treatments with certain pharmacological agents.[38] Teratogens often elicit their effects through specific signaling pathways. Transgenic mice with targeted mutations in receptors or other components of signaling pathways are available and potentially can be used in conjunction with teratogen exposure to assess the involvement of these pathways in teratogenesis. For example, targeted mutants have been made for components of retinoic acid signaling, including cellular binding proteins[39] and various nuclear receptors.[40,41] Exposure of these various mutants, or combinations of mutants, to retinoids may provide insight into the mechanisms of retinoid teratogenicity. There are many other examples of targeted mutations which result in cellular and structural abnormalities similar to those seen in response to teratogen exposure. Not only does the resulting phenotype provide information with regard to the importance and function of the gene, but it also may provide insight into the pathogenesis and mechanisms involved in teratogenesis.

2. Diverse Genetic Manipulations

Insertional inactivation of genes has been the most frequently used genetic alteration in gene targeting, but a full range of genetic alterations are possible, including point mutations, large or small deletions of DNA, gene duplications or inversions, or the creation of fusion genes.[42] Although less well characterized than insertional mutagenesis, these types of gene manipulations no doubt will develop into powerful new approaches to studying gene regulation and function. An example of this is seen with the generation of a transgenic mouse model to study cystic fibrosis. Cystic fibrosis is a common genetic disorder for which approximately 1 in 15 Caucasians is a carrier of a mutated CFTR gene. The most common form of the disease results from a specific 3-base-pair deletion in the CFTR gene. Recently, homologous recombination techniques in ES cells were used to created transgenic mice with the same 3-base-pair deletion.[43] These mice go on to develop features indicative of cystic fibrosis in humans and can be used to study aspects of the disease specifically related to this mutation.

IV. When Two Approaches Are Better Than One

Often genetically manipulated mice can be used in combination with one another to address specific questions. This strategy is abundant in the field of immunology where transgenic mice harboring various gene disruptions in cytokines or cell surface receptors are mated with one another to assess the relative involvement of various signaling pathways in the development and function of lymphoid cells. The sophistication of such matings is amplified by the combination of gene-targeted mutations with transgenic mice ectopically expressing signaling components.

A. Tissue- and Site-Specific Gene Deletion

It is not uncommon for null mutations generated by gene targeting to result in embryonic or fetal lethality. Whereas this demonstrates the importance of a gene during development, it prevents investigation of its function during later stages of development or in adult tissues. This reveals a caveat of standard gene targeting approaches. New technology combining both zygote microinjection and gene targeting in ES cells now provides a means of creating conditional or tissue-specific gene disruptions.[44a] One approach to this is based on the Cre-*lox*P recombination system of bacteriophage P1. Cre is a recombinase which recognizes specific short sequences of phage DNA called *lox*P sites and removes DNA sequences between these sites. The first step in creating a tissue-specific gene deletion is to create a transgenic mouse in which *lox*P sequences flank the sequences of a gene which are to be deleted. This is done using gene targeting in ES cells. Next, zygote microinjection is used to create transgenic mice which express the Cre recombinase only in the specific cells in which the targeted deletion is desired. This is accomplished using tissue-specific DNA regulatory elements to target the expression of Cre. When these mice are mated onto the background containing *lox*P sequences, the Cre recombinase will recognize the *lox*P sequences in the tissues where it is expressed and remove the DNA between the sites, thus creating cell type-specific deletion mutations. This approach has been demonstrated successfully for the T-cell-specific deletion of the DNA polymerase β gene,[44b,45] and no doubt is currently being pursued for the tissue-specific inactivation of many other genes.

B. Tissue-Specific Genetic Rescues

Another approach which combines both gene targeting and zygote microinjection to circumvent phenotypic bottlenecks is the genetic rescue of null mutations through tissue-specific replacement of gene products. An example of this is seen in studies designed to produce a mouse model for cystic fibrosis. Mice carrying a targeted disruption in the CFTR gene die from an intestinal phenotype which limits the use of these animals as models for the lung phenotype associated with cystic fibrosis in humans.[46] To circumvent this problem, investigators mated CFTR null mice with transgenic mice expressing a normal CFTR minigene under the control of intestinal DNA regulatory elements.[47] They were able to successfully rescue these animals from the intestinal phenotype, and, since the CFTR minigene was expressed only in the intestine, these animals could now be utilized as models for the airway phenotype common to humans. Another example of this approach is demonstrated in the efforts to develop mice deficient in the adenosine deaminase (ADA) gene, with the intent of generating mouse models for the severe combined immunodeficiency which occurs in ADA deficient humans. When the murine ADA gene was disrupted by gene targeting, mice homozygous for the null ADA allele died perinatally,[48,49] suggesting that ADA is essential for prenatal development in mice, but also preventing analysis of the postnatal immune system. ADA is highly expressed in the murine placenta during fetal stages of development.[50] ADA-deficient fetuses were rescued from perinatal lethality by genetically restoring ADA to the placenta.[51] This was accomplished by expressing an ADA minigene in the placenta of transgenic mice under the control of placental-specific DNA regulatory elements isolated from the murine ADA gene itself.[52] Mating this minigene onto the ADA-deficient background restored placental expression and rescued ADA deficient fetuses from perinatal lethality.[51] These rescued mice went on to exhibit metabolic and immunologic features relevant to those seen in ADA-deficient humans and can be used as models to understand the metabolic mechanisms of this disease as well as to advance the treatment of such genetic disorders.[53]

V. Availability of Transgenic Mice

Transgenic mouse technology has greatly enhanced our understanding of how genes are expressed and how gene products function in complex biological settings. The ability to manipulate gene expression in a specific manner has opened the door to countless opportunities for the development of mouse animal models for studying mechanisms and developing treatments for genetic disorders and diseases. The use of transgenic technology need not be limited to those laboratories which are equipped to generate transgenic mice. Many of the mutations and transgenics generated by other investigators are available for general scientific use through collaborative or commercial opportunities. A powerful tool for finding out what transgenic animals are available is the "Mouse and Rat Research Home Page", which can be accessed through the World Wide Web. On this home page can be found databases such as TBASE, which contains comprehensive and current listings of all transgenic animals and targeted mutations generated internationally. Also found is information regarding the availability and acquisition of transgenic mice from Jackson Laboratories, National Institutes of Health, and Oak Ridge National Laboratories, as well as arrangements necessary for having specific mice made. The locator internet addresses for this home page and other important locators for various transgenic mouse facets are listed below. The practical feasibility of transgenic mouse research has advanced to the stage where nearly every area of mammalian research can and should take advantage of genetically engineered mice.

Mouse and Rat Research Home Page

http://www.cco.caltech.edu/~mercer/htmls/rodent_page.html

TBASE

http://www.gdb.org/dan/tbase/tbase.html

Oak Ridge National Laboratory, Transgenic and Targeted Mutant Animal Database

http://www.ornl.gov/techresources/trans/hmepg.html

Gene Knockout Database

http://www.bayanet.com/bioscience/knockout/knochome.htm

NetVet-Rodent Home Page

http://netvet.wustl.edu/rodents.htm

JAX IMR, The Jackson Laboratory, Induced Mutant Resource

http://lena.jax.org/resources/documents/imr/

The Jackson Laboratory Microinjection Service

http://www.jax.org/resources/documents/transgenics/tg.html

References

1. Gordon, J.W., Scangos, G.A., Plotkin, D.J., Barbosa, J.A., and Ruddle, F.H., Genetic transformation of mouse embryos by microinjection of purified DNA, *Proc. Natl. Acad. Sci. U.S.A.*, 77, 7380, 1980.
2. Grosveld, F. and Kollias, G., Eds., *Transgenic Animals,* Academic Press, New York, 1992.
3. Hogan, B., Beddington, R., Castantini, F., and Lacy, E., Eds., *Manipulating the Mouse Embryo: A Laboratory Manual,* 2nd ed., Cold Spring Harbor Laboratory, New York, 1994.
4. Wassarman, P.M. and DePamphilis, M.L., Eds., *Guide to Techniques in Mouse Development,* Academic Press, New York, 1993.
5. Joyner, A.L., Ed., *Gene Targeting: A Practical Approach,* Oxford University Press, New York, 1993.
6. Evans, M.J. and Kaufman, M.H., Establishment in culture of pluripotential cells from mouse embryos, *Nature,* 292, 154, 1981.
7. Martin, G.R., Isolation of a pluripotent cell line from early mouse embryos cultured in medium conditioned by teratocarcinoma stem cells, *Proc. Natl. Acad. Sci. U.S.A.,* 78, 7634, 1981.
8. Bradley, A., Evans, M., Kaufman, M.H., and Robertson, E., Formation of germ-line chimeras from embryo-derived teratocarcinoma cell lines, *Nature,* 309, 255, 1984.
9. Robertson, E., Bradley, A., Kuehn, M., and Evans, M., Germ-line transmission of a gene introduced into cultured pluripotential cells by retroviral vector, *Nature,* 323, 445, 1986.
10. Grosveld, F., van Assendelft, G.B., Greaves, D.R., and Kollias, G., Position-independent, high-level expression of the human β-globin gene in transgenic mice, *Cell,* 51, 975, 1987.
11. McKnight, R.A., Shamay, A., Sankaran, L., Wall, R.J., and Hennighausen, L., Matrix-attachment regions can impart position-independent regulation of a tissue-specific gene in transgenic mice, *Proc. Natl. Acad. Sci. U.S.A.,* 89, 6943, 1992.
12. Choi, Y., Lee, I., and Ross, S.R., Requirement for the simian virus 40 small tumor antigen in tumorigenesis in transgenic mice, *Mol. Cell Biol.,* 8, 3382, 1988.
13. Andres, A.C., van der Valk, M., Schonenberger, C.A., Fluckiger, F., LeMeur, M., Gerlinger, P., and Groner, B., *Ha-ras* and *c-myc* oncogene expression interferes with morphological and functional differentiation of mammary epithelial cells in single and double transgenic mice, *Genes Dev.,* 2, 1486, 1988.
14. Thomas, H. and Balkwill, F., Assessing new anti-tumor agents and strategies in oncogene transgenic mice, *Cancer Metastasis Rev.,* 14, 97, 1995.
15. Cardiff, R.D. and Muller, W.J., Transgenic mouse models of mammary tumorigenesis, *Cancer Surv.,* 16, 97, 1993.
16. Greaves, D.R., Fraser, P., Vidal, M.A., Hedges, M.J., Ropers, D., Luzzatto, L., and Grosveld, F., A transgenic mouse model of sickle cell disorder, *Nature,* 343, 183, 1990.

17. Ryan, T.M., Townes, T.M., Reilly, M.P., Asakura, T., Palmiter, R.D., Brinster, R.L., Behringer, R.R., Human sickle hemoglobin in transgenic mice, *Science,* 247, 566, 1990.
18. Quon, D., Catalano, W.R., Scardina, J.M., Murakami, K., and Cordell, B., Formation of β-amyloid protein deposits in brains of transgenic mice, *Nature,* 352, 239, 1991.
19. Westphal, H., Transgenic mammals and biotechnology, *FASEB,* 3, 117, 1989.
20. Pool, R., Molecular biology lies down with the lamb, *Science,* 249, 124, 1990.
21. Archibald, A.L., McClenaghan, M., Hornsey, V., Simons, J.P., and Clark, A.J., High-level expression of biologically active human α_1-antitrypsin in the milk of transgenic mice, *Proc. Natl. Acad. Sci. U.S.A.,* 87, 5178, 1990.
22. Lonberg, N., Taylor, L.D., Harding, F.A., Trounstine, M., Higgins, K.M., Schramm, S. R., Kuo, C.C., Mashayekh, R., Wymore, K., McCabe, J.G., Munoz-O'Regan, D., O'Donnell, S.L., Lapachet, E.S.G., Bengoechea, T., Fishwild, D.M., Carmack, C.E., Kay, R.M., and Huszar, D., Antigen-specific human antibodies from mice comprising four distinct genetic modifications, *Nature,* 368, 856, 1994.
23. Han, L., Yun, J.S., and Wagner, T.E., Inhibition of Moloney murine leukemia virus-induced leukemia in transgenic mice expressing antisense RNA complementary to the retroviral packaging sequences, *Proc. Natl. Acad. Sci. U.S.A.,* 88, 4313, 1991.
24. Imakado, S., Bickenbach, J.R., Bundman, D.S., Rothnagel, J.A., Attar, P.S., Wang, X. J., Walczak, V.R., Wisniewski, S., Pote, J., Gordon, J.S., Heyman, R.A., Evans, R.M., and Roop, D.R., Targeting expression of a dominant-negative retinoic acid receptor mutant in the epidermis of transgenic mice results in loss of barrier function, *Genes Dev.,* 9, 317, 1995.
25. Chang, P.Y., Benecke, H., Marchand-Brustel, Y.L., Lawitts, J., and Moller, D.E., Expression of a dominant-negative mutant human insulin receptor in the muscle of transgenic mice, *J. Biol. Chem.,* 269, 16034, 1994.
26. Yamaizumi, M., Mekada, E., Uchida, T., and Okada, Y., One molecule of diphtheria toxin fragment A introduced into a cell can kill the cell, *Cell,* 15, 245, 1978.
27. Breitman, M.L., Rombola, H., Maxwell, I.H., Klintworth, G.K., and Bernstein, A., Genetic ablation in transgenic mice with an attenuated diphtheria toxin A gene, *Mol. Cell Biol.,* 10, 474, 1990.
28. Breitman, M.L., Clapoff, S., Rossant, J., Tsui, L.C., Glode, L.M., Maxwell, I.H., and Bernstein, A., Genetic ablation: targeted expression of a toxin gene causes microphthalmia in transgenic mice, *Science,* 238, 1563, 1987.
29. Palmiter, R.D., Behringer, R.R., Quaife, C.J., Maxwell, F., Maxwell, I.H., and Brinster, R. L., Cell lineage ablation in transgenic mice by cell-specific expression of a toxin gene, *Cell,* 50, 435, 1987.
30. Behringer, R.R., Mathews, L.S., Palmiter, R.D., and Brinster, R.L., Dwarf mice produced by genetic ablation of growth hormone-expressing cells, *Genes Dev.,* 2, 453, 1988.
31. Borrelli, E., Heyman, R., Hsi, M., and Evans, R.M., Targeting of an inducible toxic phenotype in animal cells, *Proc. Natl. Acad. Sci. U.S.A.,* 85, 7572, 1988.

32a. Heyman, R.A., Borrelli, E., Lesley, J., Anderson, D., Richman, D.D., Baird, S.M., Hyman, R., and Evans, R.M., Thymidine kinase obliteration: creation of transgenic mice with controlled immune deficiency, *Proc. Natl. Acad. Sci. U.S.A.*, 86, 2698, 1989.
32b. No, D., Yao, T.-P., and Evans, R.M., Ecdysone-inducible gene expression in mammalian cells and transgenic mice, *Proc. N. H. Acad. Sci.*, 93, 3346, 1996.
33. Bradley, A., Embryonic stem cells, proliferation and differentiation, *Curr. Opin. Cell Biol.*, 2, 1013, 1990.
34. Brandon, E.P., Idzerda, R.L., and McKnight, G.S., Targeting the mouse genome: a compendium of knockouts (part I), *Curr. Biol.*, 5, 625, 1995.
35a. Nicol, C.J., Harrison, M.L., Laposa, R.R., Gimelshtein, I.L., and Wells, P.G., A teratologic suppressor role for p53 in benzo[a]pyrene-treated transgenic p53-deficient mice, *Nat. Genet.*, 10, 181, 1995.
35b. Wubah, J.A., Ibrahim, M.M., Gao, X., Nguyen, D., Pisano, M.M., and Knudsen, T.B., Teratogen-induced eye defects mediated by p53-dependent apoptosis, *Curr. Biol.*, 6, 60, 1996.
36. Wang, Z.Q., Auer, B., Stingl, L., Berghammer, H., Haidacher, D., Schweiger, M., and Wagner, E.F., Mice lacking ADPRT and poly(ADP-ribosyl)ation develop normally but are susceptible to skin disease, *Genes Dev.*, 9, 509, 1995.
37. McWhir, J., Selfridge, J., Harrison, D.J., Squires, S., and Melton, D.W., Mice with DNA repair gene (*ERCC-1*) deficiency have elevated levels of p53, liver nuclear abnormalities and die before weaning, *Nat. Genet.*, 5, 217, 1993.
38. Schinkel, A.H., Smit, J.J. M., van Tellingen, O., Beijnen, J.H., Wagenaar, E., van Deemter, L., Mol, O.A.A.M., van der Valk, M.A., Robanus-Manndag, E.C., te Riele, H.P.J., Berns, A.J.M., and Borst, P., Disruption of the mouse *mdr1a* p-glycoprotein gene leads to a deficiency in the blood-brain barrier and to increased sensitivity to drugs, *Cell*, 77, 491, 1994.
39. Gorry, P., Lufkin, T., Dierich, A., Rochette-Egly, C., Decimo, D., Dolle, P., Mark, M., Durand, B., and Chambon, P., The cellular retinoic acid binding protein I is dispensable, *Proc. Natl. Acad. Sci. U.S.A.*, 91, 9032, 1994.
40. Lufkin, T., Lohnes, D., Mark, M., Dierich, A., Gorry, P., Gaub, M.P., Lemeur, M., and Chambon, P., High postnatal lethality and testis degeneration in retinoic acid receptor alpha mutant mice, *Proc. Natl. Acad. Sci. U.S.A.*, 90, 7225, 1993.
41. Kastner, P., Grondona, J.M., Mark, M., Gransmuller, A., LeMeur, M., Decimo, D., Vonesch, J-L., Dolle, P., and Chambon, P., Genetic analysis of RXRα developmental function: convergence of RXR and RAR signaling pathways in heart and eye morphogenesis, *Cell*, 78, 987, 1994.
42. Bronson, S.K. and Smithies, O., Altering mice by homologous recombination using embryonic stem cells, *J. Biol. Chem.*, 269, 27155, 1994.
43. Colledge, W.H., Abella, B.S., Southern, K.W., Ratcliff, R., Jiang, C., Cheng, S.H., MacVinish, L.J., Anderson, J.R. Cuthbert, A.W., and Evans, M.J., Generation and characterization of a ΔF508 cystic fibrosis mouse model, *Nat. Genet.*, 10, 445, 1995.
44a. Ezzell, C., Murine mania: researchers devise a host of new techniques for making knockout mice, *J. Natl. Inst. Health Res.*, 8, 33, 1996.

44b. Gu, H., Marth, J.D., Orban, P.C., Mossmann, H., and Rajewsky, K., Deletion of a DNA polymerase β gene segment in T cells using cell type-specific gene targeting, *Science,* 265, 103, 1994.
45. Kuhn, R., Schwenk, F., Aguet, M., and Rajewsky, K., Inducible gene targeting in mice, *Science,* 269, 1427, 1995.
46. Ratcliff, R., Evans, M.J., Cuthbert, A.W., MacVinish, L.J., Foster, D., Anderson, J.R., and Colledge, W.H., Production of a severe cystic fibrosis mutation in mice by gene targeting, *Nat. Genet.,* 4, 35, 1993.
47. Zhou, L., Dey, C.R., Wert, S.E., DuVall, D., Frizzell, R.A., and Whitsett, J.A., Correction of lethal intestinal defect in a mouse model of cystic fibrosis by human CFTR, *Science,* 266, 1705, 1994.
48. Wakamiya, M., Blackburn, M.R., Jurecic, R., McArthur, M.J., Geske, R.S., Cartwright, J., Jr., Mitani, K., Vaishnav, S., Belmont, J.W., Kellems, R.E., Finegold, M.J., Montgomery, C.A., Jr., Bradley, A., and Caskey, C.T., Disruption of the adenosine deaminase gene causes hepatocellular impairment and perinatal lethality in mice, *Proc. Natl. Acad. Sci. U.S.A.,* 92, 3673, 1995.
49. Migchielsen, A.A., Breuer, M.L., van Roon, M.A., te Riele, H., Zurcher, C., Ossendorp, F., Toutain, S., Hershfield, M.S., Berns, A., and Valerio, D., Adenosine-deaminase-deficient mice die perinatally and exhibit liver-cell degeneration, atelectasis and small intestine cell death, *Nat. Genet.,* 10, 279, 1995.
50. Knudsen, T.B., Blackburn, M.R., Chinsky, J.M., Airhart, M.J., and Kellems, R.E., Ontogeny of adenosine deaminase in the mouse decidua and placenta: immunolocalization and embryo transfer studies, *Biol. Reprod.,* 44, 171, 1991.
51. Blackburn, M.R., Wakamiya, M., Caskey, C.T., and Kellems, R.E., Tissue-specific rescue suggests that placental adenosine deaminase is important for fetal development in mice, *J. Biol. Chem.,* 270,23891, 1995.
52. Winston, J.H., Hanten, G.R., Overbeek, P.A., and Kellems, R.E., 5′ Flanking sequences of the murine adenosine deaminase gene direct expression of a reporter gene to specific prenatal and postnatal tissues in transgenic mice, *J. Biol. Chem.,* 267, 13472, 1992.
53. Blackburn, M.R., Datta, S.K., Wakamiya, M., and Vastabedian, B.S., Metabolic and immunologic consequences of limited adenosine deaminase expression in mice, *J. Biol. Chem.* (in press).

Chapter 4

The Use of Antisense Oligonucleotide Technology and Mouse Whole-Embryo Culture to Study Gene Function During Organogenesis

Linda Foerst Potts and T.W. Sadler

Contents

0-8493-3342-3/97/$0.00+$.50

I. Introduction

In the past few years, new molecular biology techniques have been applied to study the function and expression patterns of particular genes in embryonic development. One such technique, targeted gene disruption, allows researchers to study the function of a particular gene by genetically engineering null mutant mice. This approach has been useful for identifying the roles of zygotic genes essential for embryogenesis and, in some cases, the functions of genes during the period of embryogenesis. However, when developmental genes play essential roles at several stages in embryogenesis, this approach may be limited to delineation of the initial function of a gene since mutant embryos blocked at an early stage may not progress to later stages. Furthermore, in the null mutant model the resulting phenotype often is the result of a complex interplay between the loss of a gene and the embryo's attempt to circumvent that loss. Thus, interpretation of the results from these mice is often complicated. Therefore, an alternative technique, antisense oligonucleotide technology, has been developed that permits targeted gene disruptions that not only are more economical and less time-consuming than those obtained by engineering null mutant mice, but also can be used to block a gene's function at specific stages of development.

The term "antisense" describes the interaction between oligodeoxynucleotides (ODNs) complementary to (sense) pre-mRNA or mRNA molecules that inhibit the production of the protein product.[1] Usually the oligonucleotide probes are short sequences of DNA targeted to a specific region of mRNA. When using antisense technology in conjunction with mouse whole-embryo culture, antisense oligonucleotides are injected into the amniotic cavity of mouse embryos at early somite stages of embryogenesis. ODNs are taken up by embryonic cells by endocytosis and, once inside the cell, exert their effects by several mechanisms (Figure 4.1).[2,3] The major mechanism is through the action of RNase H. After the ODN binds to the mRNA, an endogenous ribonuclease, RNase H, which recognizes RNA-DNA hybrids, cleaves the RNA, prohibiting translation.[4] ODNs bound to the 5′ untranslated region of mRNA also can cause steric hindrance and prevent binding and sliding of the 40S ribosomal subunit and/or the association of protein factors involved in the initiation of translation.[5] Finally, ODNs can interfere with transcription by

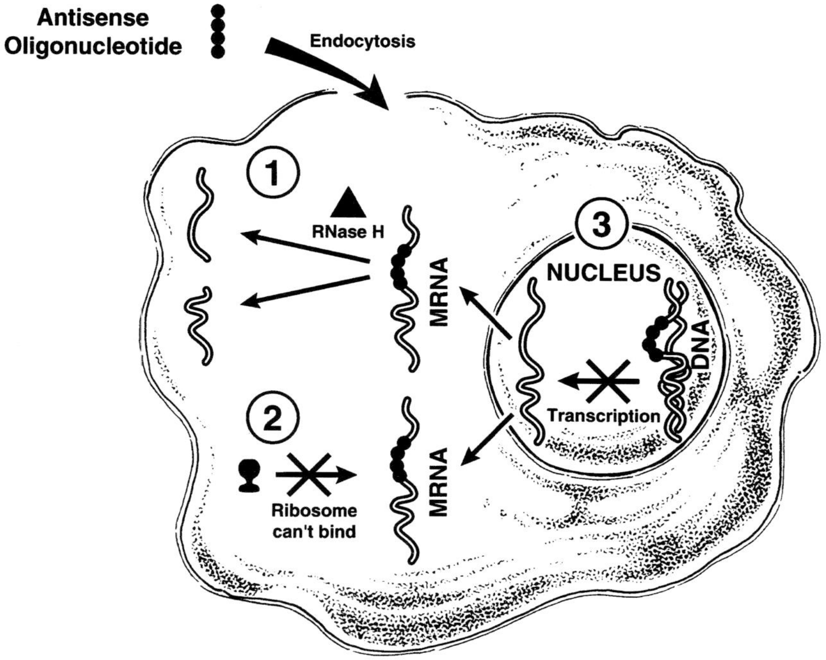

FIGURE 4.1
Mechanisms of antisense oligonucleotides (ODNs): Antisense ODNs are taken up by the cell by endocytosis and can exert their effects by several mechanisms. **(1)** Once the ODN binds to the mRNA, RNase H, an enzyme which recognizes RNA-DNA hybrids, cleaves the RNA portion, preventing translation. **(2)** ODN binding to the mRNA causes steric hinderance, which prevents binding of the ribosome. **(3)** ODNs prevent transcription by forming a triplex helix with DNA.

binding to the DNA double helix, forming a triplex, or by hybridizing with the intron-exon junction to inhibit splicing.[6]

II. Scope and Limitiations

The use of antisense ODNs to inhibit gene expression has the potential to contribute to the fields of cancer and developmental biology. The development of effective antisense techniques will allow researchers to address questions about gene function and may also lead to new gene therapies for many types of human diseases. However, while antisense ODNs have the ability to act in a sequence-specific manner, caution is required when conducting and interpreting antisense studies, since many biological effects of ODNs may not be attributable to specific effects on targeted mRNAs. For example, charged ODNs are polyanions and may bind and sequester heparin binding growth factors. Like heparin, phosphorothioated ODNs, such as SdD28, appear to block the binding of bFGF and PDGF to receptors on the surface of 3T3 fibroblasts and other cells.[7] Phosphorothioated ODNs may also exhibit sequence-selective nonantisense effects. For example, ODNs containing four contiguous guanosine residues may be antiproliferative in an otherwise sequence-independent manner.[8]

Therefore, when conducting antisense studies, it is crucial that experiments be well designed such that reliable conclusions concerning results can be made. For example, the target sequence must be selected carefully and should be specific for the mRNA to be targeted. Due to the secondary structure of RNA, ODNs may not always bind readily.[9] Therefore, it may be necessary to try several ODNs to different regions of the targeted mRNA to obtain an effective probe.[10] Also, nuclease-resistant ODNs should be used. In this regard, phosphorothioated ODNs are more stable as compared to phosphodiester linkages, which are rapidly degraded in most cells.[11] It is also imperative to demonstrate a decrease in either the target mRNA or protein as part of the antisense mechanism by semiquantitative PCR or immunohistochemistry. Better ODN delivery systems need to be tested. Recently, cationic liposomes have been developed in which the surface of these liposomes is positively charged and is thus attracted to the phosphate backbone of DNA in ODNs, as well as the negatively charged surface of the cell membrane. Lipid-DNA complexes are formed and have been shown to more effectively deliver ODNs to the cytoplasm.[12-14] Finally, an acceptable number of controls should be performed to demonstrate antisense specificity.[15]

Five types of controls commonly used with ODNs are (1) phosphate-buffered saline (PBS; vehicle); (2) sense ODNs; (3) mismatched ODNs, in which 2 to 4 bases are changed in the antisense oligonucleotide sequence; (4) scrambled ODNs, in which the same base composition of the antisense oligonucleotide is rearranged; and (5) multiple antisense ODNs to different regions of the targeted mRNA, which should produce the same phenotypes.[16] A sense control may present problems, namely the "anti-gene" strategy in which the sense ODNs may be able to form a triple helix with DNA to inhibit transcription.[5] Mismatched controls in which 2 to 4 bases of a 20-base oligo are changed are the most specific type of control, but they may not always be reliable in that, in some cases, only four bases are required to confer specificity and to activate RNase H.[5] Thus, it may be possible that as few as four unchanged bases in the mismatch control are still able to elicit an effect. Scrambled controls sometimes present problems because their sequence may match that of an unknown gene; therefore, more than one type of control should be employed when conducting antisense studies.

III. Advantages and Disadvantages

Although antisense technology presents some difficulties, the technique does provide some unique advantages as a new approach for developmental biologists to study gene function. Previously, genes thought to play critical roles in embryogenesis were targeted and the function assessed through site-directed mutagenesis and the generation of null mutant mice. This approach has resulted in a multitude of interesting models involving knockouts of homeobox genes, proto-oncogenes, and others. However, the technique and the models it creates have several disadvantages in that the technique is expensive and time consuming. In contrast, using whole-embryo culture in conjunction with antisense technology provides rapid results, usually in 24 to 48 hours vs. 1 to 2 years with transgenic mice and targeted knockouts. In addition, embryos can be staged precisely prior to experimentation, which is important because a gene of interest may be expressed at different times during embryogenesis and therefore may have different functions during each of these times. Mouse whole-embryo culture is a method that provides a way to investigate the function of the same gene during different developmental periods. For example, the gene of interest may be expressed during gastrulation but also may be functional and expressed in regions where neural crest cells contribute to facial mesenchyme. With the gene knockout technique, the latter function might be missed, since disruption of the gene during gastrulation may result in death of the embryo. However, both roles for the gene can be investigated using the antisense approach, since one can inject antisense ODNs at both of these stages of development, independently.

Another disadvantage of knockout mice is that, in most cases, only animals homozygous for the null alleles exhibit an abnormal phenotype, and these animals usually do not survive to breeeding age. Followup studies designed to determine the earliest cellular events (cell proliferation, cell migration, cell death, etc.) that produce the phenotype are difficult to perform, because only 25% of offspring from a mating of heterozygotes will be abnormal and they cannot be identified prior to exhibiting the phenotype except by genetic markers. Thus, studies to investigate cellular mechanisms responsible for the phenotype often involve using all the embryos from a litter for an experiment, collecting data, and then hoping that 25% will show a difference from the others. Such an approach is time consuming and may be limited by the amounts of tissue available for both cell study and genotyping. Antisense studies conducted by our laboratory with several different genes have generated a phenotype in about 60 to 80% of the embryos injected, making followup studies easier and less time-consuming.

Another advantage to using antisense technology with mouse whole-embryo culture as compared to the knockout approach is that the study of gene compensation can be addressed readily. Phenotypes in null mutant mice are not simply the result of the absence of a gene but are also dependent on the ability of other genes or groups of genes to compensate for that loss.[17] In some cases, redundant family members may assume the function of the deleted member or cells may make decisions which affect groups of genes important in specific developmental pathways. Thus, given that the gene has been disrupted since the time of fertilization in null mutant mice, there may be more time for the organism to compensate or make up for the loss of a gene as compared to disrupting a gene during a specific stage of development using the antisense approach. In addition, functional redundancy between genes is difficult to test using the targeted disruption approach, as it requires a significant amount of time to breed the mice that are missing two or more genes. However, fuctional redundancy can be readily addressed using antisense technology with the mouse whole-embryo culture system by knocking out two or more related genes, simultaneously.

In summary, advantages of the antisense approach follow:

1. The antisense approach is economical and results can be obtained in 24 to 48 hours.
2. Genes can be targeted during specific stages of development to assess their function.
3. Followup studies at the cellular level can be conducted more readily and rapidly.
4. The opportunity for compensation is reduced, as genes are targeted during specific stages of development.
5. Multiple genes can be tested simultaneously to test for functional redundancy.

A disadvantage to using the antisense approach with mouse whole embryo culture is that only a limited number of embryonic stages can be studied, because embryos can only be grown from gastrulation through early phases of organogenesis. Therefore, embryos cannot be maintained to fetal stages to assess the final phenotypic results of a gene disruption. This limitation, however, can be partially circumvented by the ability to grow various cells, tissues, or organs using other culture techniques for additional periods of time.

IV. Methods

A. Culture Technique

To conduct antisense studies using mouse whole-embryo culture, one must master and feel confident using the whole-embryo culture technique. It is crucial that embryos can be grown normally in culture before attempting any antisense studies. For the preparation of cultured embryos, random-bred ICR strain mice (Harlan Laboratories; Indianapolis, IN) have been employed in previous studies and are a hearty strain of mice not known to be genetically deficient. Males are mated overnight and examined for the presence of a sperm plug the following morning, which is designated as gestational day 1 (GD1) of pregnancy. At GD 9 (4 to 6 somites), embryos are prepared for culture as previously described.[18,19] Using the embryo culture technique, gastrulation (day 8) and neurulation (early somite; day 9) embryos can be maintained throughout much of the period of organogenesis. Days 8 through 12 can be

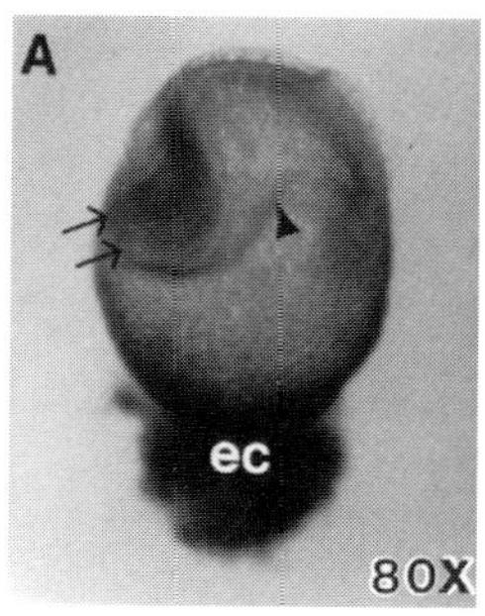

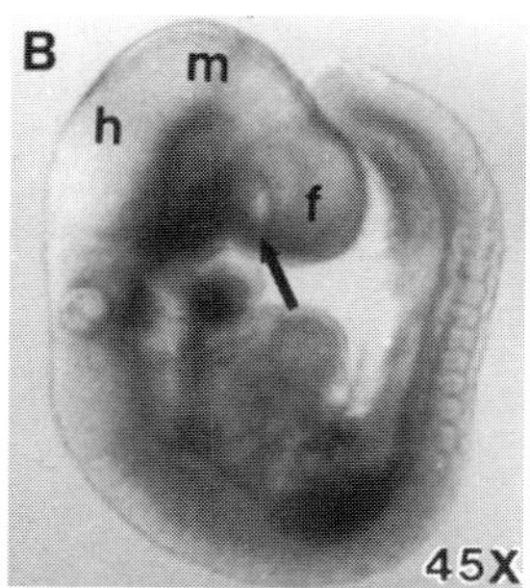

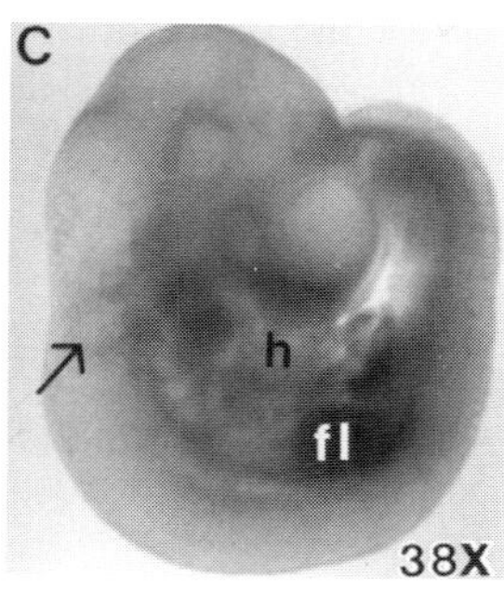

FIGURE 4.2
(A) Day 9 mouse embryo (4 to 6 somite stage) prepared for whole embryo culture showing the cranial region of the neural folds (arrows), amniotic membrane (arrowhead), and ectoplacental cone (ec). **(B)** Normal development of an embryo (22-somite stage) cultured for 24 hours. The optic vesicle is present (arrow), and the forebrain (f), midbrain (m), and hindbrain (h) regions are well segregated. **(C)** Normal development of an embryo (30-somite stage) cultured for 48 hours. By 48 hours, the forelimb bud is present (fl). The forebrain, midbrain, and hindbrain regions are well expanded and the otic vesicle (arrow), heart (h), and optic vesicle are evident, as is also seen in the 24-hour cultured embryo.

examined in most studies, coinciding with approximately the third to the fifth weeks of human development. During this time, embryos grow normally and the neural tube closes, the cardiovascular system develops, craniofacial morphology is established, and limb buds are formed (Figure 4.2). Therefore, a variety of morphological parameters are available for analysis.

Whole-embryo culture involves removing embryos from the uterine horns and placing them in Tyrodes's buffer to continue the dissection process (Figure 4.3). Then, the parietal yolk sac and Reichert's membrane are removed, leaving the visceral yolk sac, amnion, and the embryo itself intact. Culture medium consists of 50% mouse serum, 25% rat serum, and 25% Tyrode's buffer if cultures are initiated on day 8; 75% rat serum and 25% Tyrode's buffer on days 9 to 11. Embryos require a minimum of 1.0 ml of medium per 24 hours and usually are cultured separately in 2.5 ml of medium in stoppered glass flasks rotating at 30 rpm at 37°C . Embryos are gassed at 12-hour intervals throughout incubation, with increasing O_2 concentrations each day plus 5% CO_2 and the balance N_2. At 0 and 12 hours, cultures are gassed with 5% O_2, 5% CO_2, and 90% N_2; at 24 hours, with 20% O_2, 5% CO_2, and 75% N_2; and with 95% O_2 and 5% CO_2 for the remainder of the culture period. Injections of antisense ODNs can be made at any of the stages of development described. Following injection, embryos can be maintained in culture for periods of 72 hours (day 8 cultures), 48 hours (day 9 cultures), or 24 hours (day 10 and 11 cultures). At the end of the culture period, analysis of the embryos consists of a gross examination under a dissecting microscope to monitor visceral yolk sac circulation, somite number as a measure of growth, and the presence of gross structural abnomalities such as facial hypoplasia, cardiac hypertrophy, abnormalities of cardiac looping, neural tube defects, somite irregularities, etc.

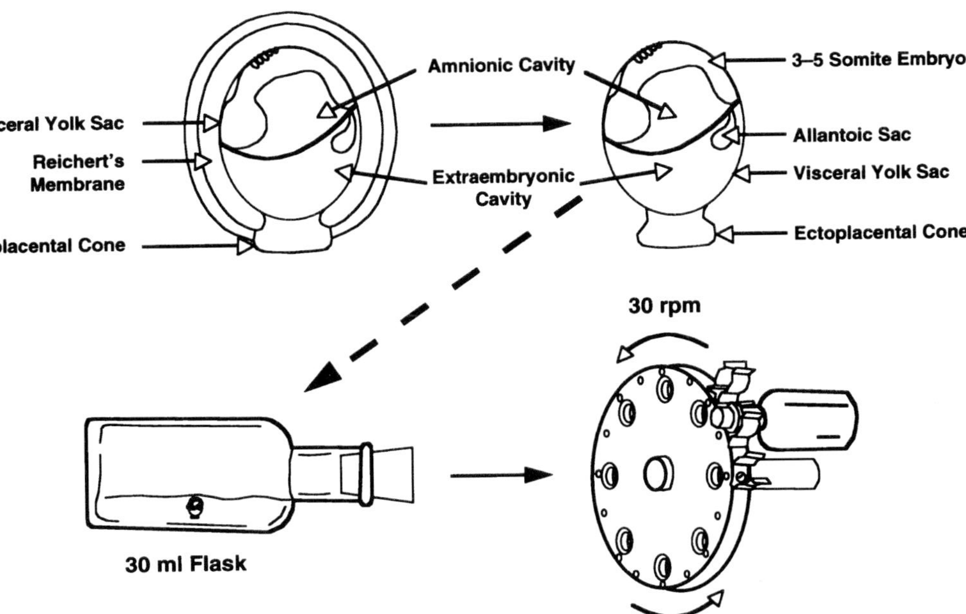

FIGURE 4.3
Mouse whole-embryo culture. Day 9 embryos are removed from the uterine horns and the parietal yolk sac, and Reichert's membrane are then removed, leaving the visceral yolk sac, amnion, and embryo intact. The embryo is placed into a 30-ml flask containing rat serum and Tyrode's buffer solution and gassed. The flasks containing embryos are then placed on a rotator wheel inside a 37°C incubator and cultured for 24 to 48 hours.

B. Oligonucleotide Synthesis and Microinjection

Several parameters should be considered before selecting antisense ODNs for the gene of interest. First, it is imperative to select an oligonucleotide that has no known significant homology to any other gene. GenBank and Blast are two well known computer programs that provide a fast, reliable method to perform this function.[20,21] Mismatch, scrambled, and sense controls also should be screened for homologies to other known genes. Second, antisense ODNs should be designed as short as possible to achieve a higher specificity. However, the ODN also should have a minimum length to avoid binding to complementary sequences that might exist in the mRNA of other genes and to achieve sufficient binding strength to have an inhibitory effect. In previous studies, most antisense ODNs shown to have an effect on translation have ranged from 12 to 20 nucleotides in length.[22,23] Successful studies conducted in our laboratory have used ODNs 20 to 21 bases in size. Third, the region of the mRNA to be targeted must also be taken into consideration. The ATG site or methionine start codon site is a common starting place. Many laboratories have found it prudent to look at the secondary structure of the mRNA, for in certain regions it may prevent longer ODNs from hybridizing;[9] however, it has been reported that antisense ODN inhibition is independent of location of the target sequence.[10] Thus, it is confusing as to why ODNs targeted to some regions are successful, whereas others are not. It may be necessary to try several ODNs targeted to different regions of the gene of interest. Fourth, the type of ODN modification chosen may play a crucial role in inhibition of translation. In this regard, ODNs are chemically modified to increase their nuclease resistance and improve their uptake by cells in culture. Initial antisense experiments used conventional DNA linked by classical phosphodiester bonds between the bases. The efficacy of this type of ODN is greatly reduced because they are easily and quickly destroyed by cellular nucleases. For example, the half-life of phosphodiester ODNs in serum can be as little as 15 minutes in some cell culture conditions.[24] However, there are several types of chemical modifications that can be made on the phosphodiester chain to make it resistant to nucleases. Two common types of modifications are phosphorothioate ODNs, in which an oxygen atom of the phosphate group is replaced by a sulfur atom, and methylphosphonate ODNs, in which an oxygen atom is replaced by a methyl group.[25] Another way of possibly improving the success rate may depend on the use of uniformly substituted or chimeric ODNs. Recent evidence has suggested that chimeric probes, in which the middle 6 to 8 bases are left unmodified, improves binding to the targeted mRNA and increases enzyme degradation of the nucleic acid, while maintaining the resistance of the probe to degradation itself.[26] Both phosphorothioated uniformly substituted ODNs, in which all bonds are modified, and chimeric probes have been used successfully in our laboratory. However, more testing needs to be performed to determine whether one type is more potent than the other.

The number of ODN synthesizing facilities has grown significantly over the last 5 years. If no local facility is available, orders can be placed to several companies throughout the U.S., and delivery can be as rapid as 24 hours. For experiments in our laboratory, phosphorothioated ODNs were synthesized at a nearby facility on an Applied Biosystems (Foster City, CA) 380 DNA synthesizer and detritylated. For antisense studies used in conjunction with mouse whole embryo culture, ODN purification is essential. Stocks are suspended in 500 μl of sterile phosphate buffered saline (PBS) and particulate matter removed by slow centrifugation. ODNs are ethanol precipitated and the pellet washed with cold ethanol and vacuum dried. ODNs then are resuspended in PBS and further purified by separation through a Sephadex G-25 column equilibrated with PBS. Yield and molarities of the purified solutions are determined by optical density and then aliquoted and stored at –20°C . ODNs are diluted in sterile PBS to the appropriate concentrations (10 to 100 μ*M*) prior to microinjection.

Once embryos are prepared for culture, microinjections are made. Microinjections involve the use of calibrated beveled glass needles (20 μ*M* diameter) that are fashioned on a pipet puller and pipet sharpener (see Appendix). The injection apparatus consists of the needle connected by silastic tubing to an air-filled syringe. Injections are performed by hand and needle tips are positioned in the amniotic cavity by penetrating the visceral yolk sac and amniotic membrane (Figure 4.4). A 100-nl volume of ODN is then dispensed into the amniotic fluid. Various concentrations of ODNs are injected to determine a dose response. In addition, embryos are injected with PBS (phosphate buffered saline) which serves as a control for the injection technique itself. At the end of culture (usually 24 to 48 hours later), embryos are stripped of their extraembryonic membranes and are staged by somite number. Gross morphology is examined by light microscopy, and developmental abnormalities and the presence of heartbeats and circulation are evaluated.

V. Gene Expression Studies

With the use of antisense technology, once a phenotype is obtained, it is absolutely crucial to determine specificity. To make this determination, more than one control (scrambled, mismatch, multiple antisense, or sense) ODNs should be employed. In addition, it is imperative that a specific effect on the protein product or the mRNA of the targeted gene be demonstrated. The effect on the protein product using immunohistochemistry is the fastest, simplest method, but many times antibodies to the protein are not available. Alternatively, demonstrating downregulation of the targeted mRNA can be accomplished using reverse transcriptase PCR or *in situ* hybridization. However, if PCR is employed, previous studies conducted in our laboratory have shown that primers should be selected to include the region of the targeted site. Since

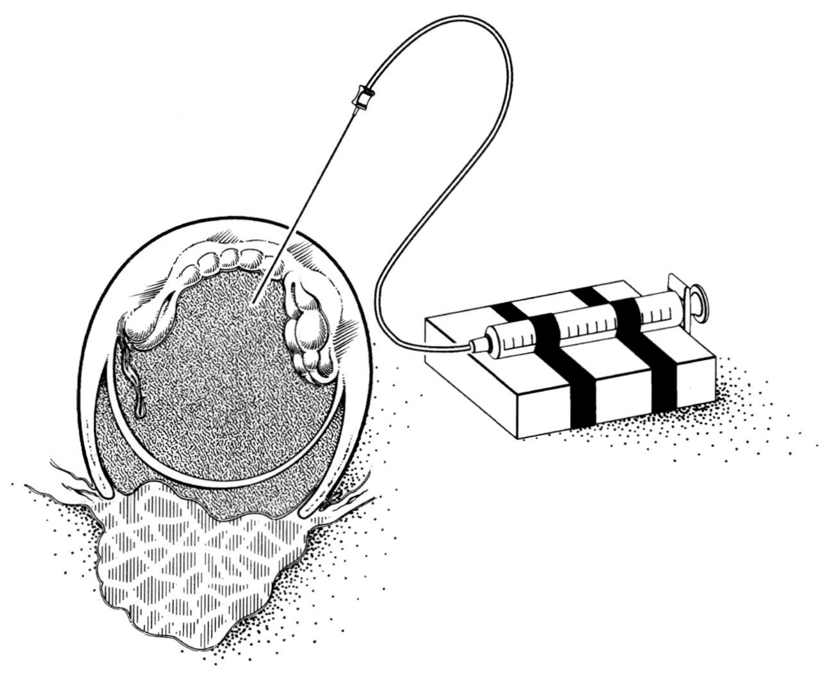

FIGURE 4.4
Schematic showing a day 9 embryo being injected with oligodeoxynucleotides. The 20-μM injection needle which is attached to silastic tubing connected to an air-filled syringe pierces the visceral yolk sac in a region devoid of embryonic tissue. Oligonucleotides dissolved in phosphate buffered saline are then dispensed into the amniotic cavity with the hand-held syringe.

RNase H cleaves only this region, two large pieces of mRNA could potentially be left that may be stable for an extended period of time. Thus, if primers are designed to amplify a region that did not encompass the targeted mRNA site, no difference in mRNA levels would be detected. Caution should be exercised when looking for quantitative effects on mRNA using *in situ* hybridization for similar reasons given that the riboprobe may be able to bind to leftover pieces of mRNA not cleaved by RNase H.

A. RNA Isolation for Semi-Quantitative PCR

Semi-quantitative polymerase chain reaction (PCR) involves comparing the amount of the targeted gene's mRNA to a reference or housekeeping gene's mRNA within the same embryo. To perform semiquantitative PCR, ideal PCR conditions first must be determined. Often, it is easier to choose the best primers and annealing conditions using one of the many readily available computer programs on the market today. Initially, experiments need to be conducted to determine amplification dynamics of the housekeeping gene (such as β-actin) and the targeted gene, using the primers and PCR conditions selected. To accomplish this standardization, a time course evaluation of PCR product sampled every 5 cycles, from 15 to 40 cycles, should be performed on both the housekeeping gene and targeted gene to determine the cycle range that exhibits linear amplification dynamics. The optimal cycle number must be selected in the linear range rather than in the plateau phase, since at higher cycle numbers amplification is nonquantitative. Also, to obtain optimal amplification, PCR conditions may have to be adjusted. In addition, a PCR time-course experiment should be conducted to determine at what time after injection of antisense ODNs a reduction in mRNA is present. Previous experiments in mouse whole-embryo culture have shown that, following injection, uptake of ODNs into embryonic cells was detectable at 45 minutes, reached significant levels by 2 hours, and peaked at 8 to 12 hours. ODNs appeared to be distributed to all cells and tissues, and significant amounts of intact probes remained at 24 hours.[27] Therefore, the time it takes the antisense ODNs to have an inhibitory effect on the mRNA may vary.

Past studies have shown mRNA downregulation to be most evident between 3 and 6 hours after injection.[27] Thus, once linear amplification dynamics and time course are determined, comparisons can be made between embryos injected with antisense ODNs, control ODNs, and PBS. The methods involved in accomplishing the above objectives are RNA isolation and PCR.

B. RNA Isolation

To isolate the RNA, day 9 embryos (4 to 6 somite stage) injected with PBS, antisense ODNs, and control ODNs are collected, and each embryo is dissected in sterile PBS to remove the extraembryonic tissues. After dissected embryos are placed into microcentrifuge tubes containing .8 ml of Tri Reagent (Molecular Research Center; Cincinnati, OH) and 8 μl of Microcarrier Gel-TR (Molecular Research Center) and homogenized, samples are placed at room temperature for 5 minutes and 200 μl of chloroform are added. Samples are then shaken vigourously, stored at room temperature for 5 minutes, and centrifuged at 12,000 g for 15 minutes at 4°C. The upper aqueous phase is placed into a fresh tube and the solution is mixed with 500 μl of isopropanol, stored for 5 minutes at room temperature, and centrifuged at 12,000 g at 4°C for 10 minutes. The supernatant is removed and the pellet washed with 1 ml of 75% ethanol and centrifuged at 7500 g for 5 minutes at 4°C. Pellets are air dried and dissolved in 20 μl diethyl pyrocarbonate (DEPC)-treated water.[28]

C. cDNA Synthesis and Polymerase Chain Reaction

1. cDNA

The following reagents are combined in the order provided using the Perkin Elmer Cetus GeneAmp RNA PCR kit: 4 μl of $MgCl_2$, 2 μl of 10× PCR Buffer II, 2 μl of each dNTP, 1 μl of RNase inhibitor, 1 μl of reverse transcriptase,

1 μl of Random Hexamers, and 3 μl of prepared embryonic RNA are added to each reaction; the reaction tubes are incubated for 10 minutes at room temperature and are overlaid with a drop of mineral oil. These samples are placed in a PCR machine (Perkin Elmer Cetus; Norwalk, CT) and are exposed to the following heating cyle: 42°C for 60 minutes, 99°C for 5 minutes, and 5°C for 5 minutes.[29]

2. Polymerase Chain Reaction

Ideal PCR conditions may vary with each gene targeted. The conditions below have been successful for more than one gene targeted in our laboratory. Of the above cDNA reaction, 8 μl are added to 8 μl of 25 μ*M* $MgCl_2$, 10 μl of 10× PCR Buffer II, 2 μl of each dNTP, and 0.5 μl of AmpliTAQ DNA polymerase. The final volume is adjusted to 100 μl with ddH_2O. Cycle conditions vary based on annealing temperature and cycle number. PCR reations for the targeted gene and the housekeeping gene are conducted simultaneously in separate tubes; 10 μl of each solution containing the targeted gene and housekeeping gene are then double-loaded on a 2% agarose gel, electrophoresed, and scanned using a densitomer. Ratios between β-actin (housekeeping gene) and the targeted gene, calculated from the density of the bands, provides a semiquantitative assessment of the effects on the gene being targeted.

Polymerase chain reaction controls consist of an untreated embryo in each experimental group that underwent RNA isolation and cDNA synthesis as described, with the exception that reverse transcriptase is omitted.

VI. Immunohistochemistry

When antibodies are available, the whole-mount immunohistochemistry staining procedure developed by Dent et al.[30] is employed to detect effects of ODNs on the protein product of targeted genes. Embryos are cultured for various periods of time following treatments and terminated as stated previously. Since it is unknown as to how much time is involved to detect a significant difference in protein levels once the mRNA of the gene of interest is disrupted, several time points might need to be examined. Usually, effects on mRNA are present in 4 to 6 hours, with effects on protein levels occurring some time later and lasting for 24 to 48 hours. Collected embryos are fixed in methanol-DMSO (4:1) overnight at 4°C and subsequently bleached by washing in methanol-DMSO-30% H_2O_2 for 5 hours at room temperature and then rehydrated through a graded methanol series (30 minutes each), with a final rinse in PBS. Embryos then are washed twice with PBSMT (2% skim milk, 0.1% Triton X-100 in PBS) and incubated overnight, shaking gently at 4°C in antibody diluted 1:1000 in PBSMT. Embryos are washed twice (1 hour each) in PBSMT at 4°C and three times with PBSMT at room temperature (1 hour each), shaking gently during all washes. Anti-rabbit-goat POD (horseradish peroxidase conjugate) is diluted 1:200 in PBSMT, and embryos are incubated overnight, shaking at 4°C. Embryos are washed five times in PBSMT, washed in PBT (0.2% bovine serum albumin, 0.1% Triton X-100 in PBS) at room temperature for 20 minutes, then placed in PBT containing 0.3 mg/ml diaminobenzidine (DAB) and 0.5% $NiCl_2$ and incubated for 30 minutes. H_2O_2 is added to 0.03%, and embryos are incubated until a color change occurs (about 10 minutes). Embryos are rinsed in PBT and then dehydrated through a graded methanol series (30 minutes each) and finally cleared in benzyl alcohol:benzyl benzoate (1:2).

VII. *In Situ* Hybridization

Embryos can be analyzed for expression of the targeted gene by whole-mount *in situ* hybridization to digoxigenin-labeled RNA probes as described previously.[31,32] Whole-mount *in situ* hybridization is appropriate for determining expression patterns of the targeted gene to verify that malformations obtained following antisense injections involve these same regions. In some cases, it also may be used qualitatively to determine a reduction in mRNA levels of the gene of interest following antisense injections. However, previous experiments have shown that it is difficult to quantitate differences in mRNA expression using *in situ* hybridization. The success of the technique depends on how much of the gene is being expressed and the extent of its downregulation. Also, as stated previously, it is possible that large pieces of targeted mRNA remain after exposure to antisense ODNs and that these pieces are able to bind to the riboprobe.

To perform whole-mount *in situ* hybridization, embryos are first dissected in lactate Ringer's solution and fixed on ice in 4% paraformaldehyde in calcium- and magnesium-free phosphate-buffered saline (PBS). Embryos are then placed in fresh fixative and fixed overnight at 4°C. Embryos are then washed three times in ice-cold PBS containing 0.1% Tween-20 (PBT) and stored at –20°C until use. (All of the following procedures are performed at room temperature with shaking on a rocker, unless noted, and the incubation time of each step is approximately 5 minutes, unless noted). Embryos are bleached in 5:1 methanol/30% hydrogen peroxide (SIGMA) for 5 to 6 hours (no mixing) followed by three 10-minute washes in methanol. Embryos are rehydrated through a graded series of methanol/PBT (75, 50, 25% and 2× PBT) and then treated with proteinase K in PBT (10 μg/ml) for 5 minutes, washed twice with glycine (2mg/ml) in PBT, and postfixed with 0.2% glutaraldehyde/4% paraformaldehyde in PBS for 20 minutes. Following three PBT washes, embryos are washed twice with hybridization buffer, and prehybridized for 1 hour at 60°C (50% foramamide, .75 *M* NaCl, 1× PE, 100 μg/ml tRNA, 0.05% heparin, 0.01% BSA, and 1% SDS). Embryos are then hybridized overnight with digoxigenin-labeled RNA probe (1 μg/ml) in the hybridization buffer at 60°C. Hybridized embryos are washed twice with solution 1 (6% 5 *M* NaCl, 1% SDS, 1× PE) for 30 minutes at 60°C, followed by two

30-minute washes of solution 1.5 (1% 5 *M* NaCl, 1% SDS, 1× PE) at 54°C, and one RNAase A plus RNAase T1 treatment for 1 hour at 37°C. The embryos are then washed for 30 minutes at 52°C with solution 2 (50% foramamide, 6% 5 *M* NaCl, 1% SDS, 1× PE) and solution 3 (50% foramamide, 3% 5 *M* NaCl, 0.1% Tween 20, 1× PE). Embryos are then washed with solution 4 (10% 5 *M* NaCl, 0.1% Tween 20, 1× PE) for 20 minutes at 70°C.

Following washes, embryos are blocked for 1 hour in 2 m*M* levamisole (0.5 mg/ml)/10% heat-inactivated (H-I) sheep serum/TBST. To prevent non-specific binding of antibody, the antidigoxigenin Fab alkaline phosphatase conjugate (Boehringer Mannheim) is diluted with H-I mouse embryonic powder and H-I sheep serum in TBST, preabsorbed for 30 minutes at 4°C, and centrifuged at 2000 to 3000 rpm for 10 minutes at 4°C. Embryos then are incubated in supernatant containing antibody overnight at 4°C. This step is followed by three 5-minute and five 45-minute washes with TBST/.5 mg/ml levamisole. Embryos then are washed twice, 20 minutes each time, with NTMT/levamisole. Coloring reactions are performed with NBT (4.5 μl/ml) and BCIP (3.5 μl/ml) in NTMT/levamisole in the dark and may require a few hours to overnight without rocking. The reaction is stopped with three washes of TBST and embryos are cleared for 1 hour in 1:1 glycerol:CMFET.

VIII. Histology and Scanning Electron Microscopy

The following techniques describe effects obtained at the cellular level following the disruption of the targeted gene. They can provide key insights into what effects have occurred on cell proliferation, cell death, cell migration, and cellular interactions that cannot be seen by the light microscope. Scanning electron microscopy (SEM), in particular, provides a three-dimensional view for determining the effects of antisense probes. A major advantage of the antisense approach is that embryos can be followed over time to determine the effects of disrupting expression of a gene. With the culture system, embryos can be monitored at hourly intervals to assess not only the earliest morphological changes produced by disruption of the gene, but also subsequent morphological alterations. Therefore, a more detailed analysis can be performed on the cellular origin of malformations produced using antisense than by the generation of null mutant mice using targeted gene disruption.

A. Histological Analysis

Randomly selected embryos injected with antisense ODNs, control ODNs, or PBS are terminated, photographed, and fixed for 1 hour in Karnovsky's fixative (2% paraformaldehyde, 2% glutaraldehyde, 0.01% calcium chloride in 0.1 *M* sodium cacodylate buffer, pH 7.3).[33] Embryos are then dehydrated through an ethanol series (twice at 10 minutes each) and rinsed twice in propylene oxide (10 minutes). Embryos are placed in a 1:1 mixture of propylene oxide and Araldite and are left overnight on an aliquot mixer. An equal volume of Araldite is added to each specimen, and mixing continues for 8 more hours. The mixture is then replaced with fresh Araldite and placed in a dessicator overnight. Specimens are embedded in Araldite and placed in a 60°C oven for 24 hours to polymerize. Blocks are sectioned at 1 μ*M* thickness with an LKB Ultramicrotome V, stained with 1% toludine blue in borax and examined with a Nikon Optiphot Light Microscope.

B. Scanning Electron Microscopy

Following termination, representative embryos are selected randomly at various time points after exposure to antisense or control ODNs. Embryos are fixed in 2.5% glutaraldehyde in Sorenson's buffer overnight and if desired may be cut in cross section or sagittally to reveal internal structures. These embryos are then postfixed in 2% OsO4 for 1 to 2 hours. Specimens are dehydrated through a graded ethanol series (two at 30 minutes each) and critically point dried. Specimens are then coated with a fine layer of gold palladium and examined with a JEOL 35B SEM.[34]

IX. Caveats and Examples of the Antisense Approach with Whole-Embryo Culture

When conducting antisense studies using mouse whole-embryo culture, the pharmacodynamics of the system must be considered. Similar to teratology studies, antisense ODNs can be compared to exposing embryos to toxic compounds in that parameters such as dose, uptake, time of exposure, and half-life all have an effect on the efficacy of the ODNs. Variations in somite stage, size of the amniotic cavity, half-life of the targeted gene's protein, and the amount of mRNA present for a targeted gene also can affect experimental outcome. Because not all embryos are affected equally, the size of the embryos and antisense ODN concentration employed must be kept constant.

Several genes have been targeted using the antisense approach in conjunction with mouse whole-embryo culture. In addition, probes targeted to over 30 different sequences have not produced any malformations and are therefore viewed as "controls" that document the specificity of the antisense approach. Targeted genes that have produced altered phenotypes include *Wnt-1*, *Wnt-3a*, *En-1*, *En-2*, *Axl*, *HNF-3β*, *Shh*, *Msx-1*, and *Msx-2*.[27,35-37] Roles for the proto-oncogenes *Wnt-1* and *Wnt-3a* in brain vesicle formation, spinal cord development, and axis formation have been documented, and these genes appear to have roles similar to the homeobox-containing genes, *engrailed-1* and *engrailed-2*.[27,35-37] Recent work has focused on other homoeobox-containing genes, *Msx-1* and *Msx-2*, which play roles in craniofacial devlopment, and on hepatocyte nuclear factor 3β (*HNF-3β*) and sonic hedgehog (*Shh*), which participate in axis formation.[38-44] Most antisense studies have resulted in approximately 50 to 80% of the embryos exhibiting an abnormal phenotype.

In situ hybridization studies have implicated *Wnt-1* and *Wnt-3a* in the development of midbrain and hindbrain structures and spinal cord regions by documenting their expression in these regions.[45] In *Wnt-1* null mutant mice, cranial defects were evident, but no spinal cord abnormalities occurred.[46] Functional compensation of *Wnt-1* deficiency by related genes was suggested to explain the absence of spinal cord defects. Recent studies in our laboratory using antisense ODNs to target both *Wnt-1* and *Wnt-3a* using whole-embryo cultures have provided a unique way to test functional redundancy. Thus, antisense attenuation of *Wnt-1* expression alone resulted in mid- and hindbrain abnormalities similar to those of the knockout mice, in addition to heart abnormalites.[27] When *Wnt-3a* was targeted, forebrain and midbrain malformations resulted, and lateral outpocketings of the neural tube occurred at the level of the forelimb bud. Dual antisense disruption of *Wnt-1* and *Wnt-3a* caused defects in all brain regions and increased the number and severity of outpocketings along the primordia of the spinal cord.[35] Therefore, these two genes may complement one another within the spinal cord region.

HNF-3β is a member of the forkhead family and is expressed in the notochord and floor plate of the neural tube during gastrulation and neurulation.[41-43] Expression of *Shh*, the homologue of the Drosophila segment polarity gene hedgehog, is similar to that of *HNF-3β*, although expression of *HNF-3β* precedes that of *Shh*.[44] Phenotypes observed after exposure of 4 to 6 somite stage embryos to antisense ODNs to *HNF-3β* and *Shh* were similar and included cranial neural tube defects (exencephaly), kinking of the spinal cord, irregular somite formation, and caudal dysgenesis. Kinking of the neural tube and irregular somite formation also were observed in *HNF-3β* knockout mice.[41,42] However, cranial defects were not observed in these animals, indicating once again that the antisense approach can uncover roles for genes that are not detected in null mutants.

In some studies, problems associated with using antisense technology occurred. For example, both sense and mismatch ODNs targeting *HNF-3β* produced phenotypes similar to those resulting from exposure to antisense probes. However, scrambled ODNs, in which the original antisense sequence was randomly reorganized but the original base composition remained unchanged, produced no abnormal phenotypes. Also, whole-mount *in situ* hybridization, employed as an alternative to semiquantitative PCR to document a decrease in *HNF-3β* mRNA, was unsuccessful, perhaps for reasons cited previously. In other studies, sense ODNs targeted to *Shh* produced the same percentage of malformations as PBS injections. These probes may have the ability to disrupt gene expression at the level of transcription by forming triple helices with DNA. In support of this hypothesis, it has been shown using PCR techniques that sense ODNs at high concentrations have specific effects on mRNA levels of the targeted gene.[27]

Other genes that have been successfully investigated using the antisense approach include *Msx-1* and *Msx-2*. *Msx-1* is expressed in mesenchymal tissue underlying the neural folds at the 5 somite stage and in neural crest cells at the tips of the neural folds. At later stages, *Msx-1* continues to be expressed in neural crest-derived mesenchyme of the facial regions and in the dental papilla, endocardial cushions of the heart, and limb bud. *Msx-2* expression occurs in overlapping and adjacent regions of *Msx-1* expression.[38-40] Approximately 65% of embryos injected with various concentrations of *Msx-1* antisense ODNs during early stages of neurulation and cultured for 24 hours exhibited abnormal development, primarily in the craniofacial region. Defects consisted of small brain vesicles, exencephaly, deficiency of maxillary and mandibular arches, and enlarged and misshapened optic vesicles (Figure 4.5). A small percentage of embryos also exhibited neural tube defects which consisted of kinks along the spinal cord region. Malformations associated with the forebrain and facial processes are similar to those observed in the *Msx-1* null mutant mice and are consistent with *Msx-1* expression patterns.[38] Although eye and spinal cord abnormalities were not present in the knockout mice, they may represent an additional function for *Msx-1* discovered by using the antisense approach.

Injections of *Msx-2* antisense ODNs and *Msx-1* + *Msx-2* antisense ODNs do not result in any new malformations compared to those produced by antisense probes to *Msx-1*, but they do increase the number and severity of brain, eye, arch, and neural tube defects. The neural tube region exhibited the most significant increase in malformations resulting in multiple kinks along the presumptive spinal cord (Figure 4.5).

Exposure of embryos to mismatch *Msx-1* ODNs reduced the number of malformed embryos to 17%, which was comparable to the incidence of defects observed in PBS controls (10%). To document the specificity of the antisense probes, RT-PCR analysis showed that *Msx-1* mRNA was reduced 6 hours after exposure to antisense ODNs as compared to mismatch and PBS controls.[47]

In combination, results from these studies demonstrate the feasibility of using the antisense approach to study gene function during embryogenesis. The similarity of some of the phenotypes generated by antisense attenuation to those observed in null mutant mice supports the reliability and specificity of antisense technology. The fact that additional malformations also occur in antisense studies as compared to the knockout mice suggests that each technique is capable of uncovering unique roles for specific genes. Differences obtained in the two approaches may be due to the timing of disruption of the expression of the targeted gene and the level of disruption that occurs. Also, knockout mice that are missing a gene from the time of fertilization may have more time to compensate for the loss of that gene than embryos exposed to antisense at the beginning of organogenesis. Therefore, both approaches should continue to be employed to provide the greatest insight into the roles of genes during development.

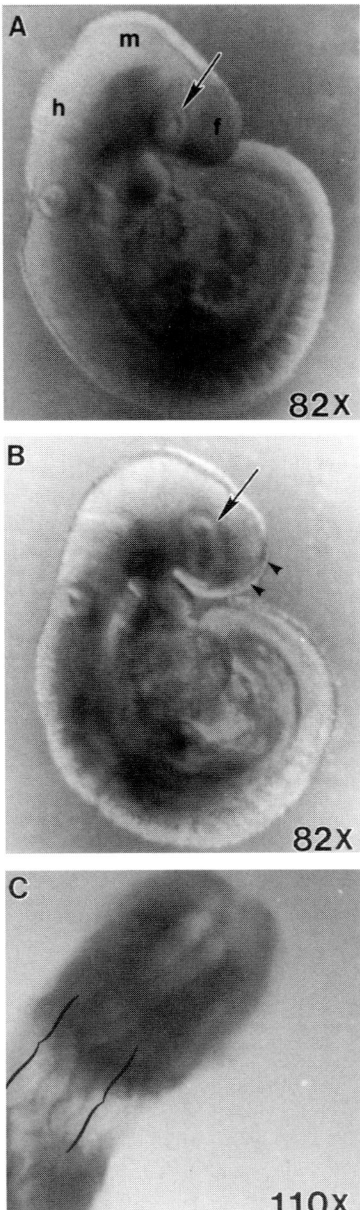

FIGURE 4.5

(A) Embryo injected with phosphate buffered saline at the 4 to 6 somite stage and cultured for 24 hours showing normal development. The optic vesicle (arrow) is visible, and forebrain (f), midbrain (m), and hindbrain (h) regions are well formed. **(B)** Embryo injected with *Msx-1* antisense oligodeoxynucleotides (60 μ*M*) at the 4 to 6 somite stage and cultured for 24 hours showing underdevelopment of the forebrain and craniofacial region (arrowheads) and expansion of the optic vesicle (arrow). **(C)** Magnified region of the dorsal neural tube of an embryo injected with *Msx-1* + *Msx-2* antisense oligodeoxynucleotides (40 μ*M* each) and cultured for 24 hours. Area between brackets demonstrates the wavy or kinked neural tube phenotype.

Acknowledgments

This work was supported by NIH grant # HD29495.

References

1. Mirabelli, C.K. and Crooke, S.T., Antisense oligonucleotides in the context of modern molecular drug discovery and development, in *Antisense Research and Applications,* Crooke, S.T. and Lebleu, B., Eds., CRC Press, Boca Raton, FL, 1993, chap. 2.
2. Loke, S.L., Stein, C.A., Zhang, X.H., Mori, K., Nakarishi, M., Subasinghe, C., Cohen, J.S., and Neckers, L.M., Characterization of oligonucleotide transport into living cells, *Proc. Natl. Acad. Sci. U.S.A.*, 86, 3474, 1989.
3. Yakubov, L.A., Deeva, E.A., Zarytova, V.F., Ivanova, E.M., Ryte, A.S., Yurchenko, L.V., and Vlassov, V.V., Mechanism of oligonucleotide uptake by cells: involvement of specific receptors?, *Proc. Natl. Acad. Sci. U.S.A.,* 86, 6454, 1989.
4. Schwab, G., Duroux, I., Chavany, C., Helene, C., and Saison-Behmoaras, E., An approach for new anticancer drugs: oncogene-targeted antisense DNA, *Ann. Oncol.,* 5 (Suppl. 4), S55, 1994.
5. Helene, C., Rational design of sequence-specific oncogene inhibitors based on antisense and antigene oligonucleotides, *Eur. J. Cancer,* 27, 1466, 1991.
6. Pierga, J.Y. and Magdelenat, H., Applications of antisense oligonucleotides in oncology, *Cell. Mol. Biol.,* 40, 237, 1994.
7. Guvakova, M.A., Yakubov, L.A., Vlodavsky, I., Tonkinson, J.L., and Stein, C.A., Phosphorothioate oligodeoxynucleotides bind to basic fibroblast growth factor, inhibit its binding to cell surface receptors, and remove it from low affinity binding sites on extracellular matrix, *J. Biol. Chem.,* 270, 2620, 1995.
8. Yaswen, P., Stampfer, M., Ghosh, K., and Cohen, J.S., Effects of sequence of thioated oligonucleotides on cultured human mammary epithelial cells, *Antisense Res. Dev.,* 3, 67, 1992.
9. Denhardt, D.T., Mechanism of action of antisense RNA. Sometime inhibition of transcription, processing, transport, or translation, *Ann. N.Y. Acad. Sci.,* 660, 70, 1992.
10. Fakler, B., Herlize, S., Amthor, B., Zenner, H.P., and Ruppersberg, J.P., Short antisense oligonucleotide-mediated inhibition is strongly dependent on oligo length and concentration, but almost independent of location of the target sequence, *J. Biol. Chem.,* 269, 1994.
11. Agrawal, S., Temsamani, J., Galbraith, W., and Tang, J., Pharmacokinetics of antisense oligonucleotides, *Clin. Pharmacokin.,* 28, 7, 1995.
12. Felgner, P.L., Gaded, T.R., Holm, M., Roman, R., Chan, H.W., Wenz, M., Northrop, J.P., Ringold, G.M., and Danielsen, M., Lipofection: a highly efficient, lipid-mediated DNA-transfection procedure, *Proc. Natl. Acad. Sci. U.S.A.,* 84, 7413, 1987.

13. Bennett, C.F., Chiang, M.Y., Chan, H., Shoemaker, J.E.E., and Mirabelli, C.K., Cationic lipids enhance cellular uptake and activity of phosphorothioate antisense oligonucleotides, *Mol. Pharmacol.,* 41, 1023, 1992.
14. Juliano, R.L. and Akhtar, S., Liposomes as a drug delivery system for antisense oligonucleotides, *Antisense Res. Dev.,* 2, 165, 1992.
15. Stein, C.A. and Krieg, A.M., Problems in interpretation of data derived from *in vitro* and *in vivo* use of antisense oligodeoxynucleotides, *Antisense Res. Dev.,* 4, 67, 1994.
16. Wagner, R.W., Gene inhibition using antisense oligodeoxynucleotides, *Nature,* 372, 333, 1994.
17. Routtenberg, A., Knockout mouse fault lines, *Nature,* 374, 314, 1995.
18. Sadler, T.W., Whole embryo culture: organogenesis of rodent embryos *in vitro*, in *Methods in Toxicology,* Vol. 1, Tyson, C.A. and Frazier, J.M., Eds., Academic Press, New York, 1993, chap. 38.
19. Sadler, T.W., Culture of early somite mouse embryos during organogenesis, *J. Embryol. Exp. Morph.,* 49, 17, 1979.
20. Burks, C., Cinkosky, M.J., Fischer, W.M., Gilna, P., Hayden, J.E.D., Keen, G., M., Kelly, M., Kristofferson, D., and Lawrence, J., GenBank, *Nucleic Acids Res. Suppl.,* 20, 2065, 1992.
21. Altschul, S.F., Gish, W., Miller, W., Meyers, E.W., Lipman, D.J., Basic local alignment search tool, *J. Mol. Biol.,* 215, 403, 1990.
22. Dash, P., Lotan, I., Knapp, M., Kandel, E.R., Goelet, P., Selective elimination of mRNAs *in vivo*: complementary oligodeoxynucleotides promote RNA degradation by an RNase H-like activity, *Proc. Natl. Acad. Sci. U.S.A.,* 84, 7896, 1987.
23. Helene, C. and Toulme, J.J., Specific regulation of gene expression by antisense, sense and antigene nucleic acids, *Biochim. Biophys. Acta.,* 1049, 99, 1990.
24. Carter, G. and Lemoine, N.R., Antisense technology for cancer therapy: does it make sense?, *Br. J. Cancer,* 67, 869, 1993.
25. Krieg, A.M., Applications of antisense oligodeoxynucleotides in immunology and autoimmunity research, *Immunomethods,* 1, 191, 1992.
26. Dagle J.M., Andracki, M.E., Devine, R.J., and Walder, J.A., Physical properties of oligonucleotides containing phosphoramidate-modified internucleoside linkages, *Nucleic Acids Res.,* 19, 1805, 1991.
27. Augustine, K., Edison, T.L., and Sadler, T.W., Antisense attenuation of *Wnt-1* and *Wnt-3a* expression in whole embryo culture reveals roles for these genes in craniofacial, spinal cord, and cardiac morphogenesis, *Dev. Genet.,* 14, 500, 1993.
28. Chomczynski, P., A reagent for the single-step simultaneous isolation of RNA, DNA and proteins from cell and tissue samples, *BioTechniques,* 1993.
29. Innis, M.A., Gelfand, D.H., Sninsky, J.J., and White, T.J., *PCR Protocols. A Guide to Methods and Applications,* Academic Press, San Diego, CA, 1990.
30. Dent, J.A., Polson, A.G., Klymkowsky, M.W., A whole-mount immunocytochemical analysis of the expression of the intermediate filament protein vimentin in Xenopus, *Development,* 105, 61, 1989.
31. Conlon, R.A., Rossant, J., Exogenous retinoic acid rapidly induces anterior ectopic expression of *Hox-2* genes *in vivo*, *Development,* 116, 357, 1992.

32. Wilkinson, D.G., Whole mount *in situ* hybridization of vertebrate embryos, in *In Situ Hybridization: A Practical Approach,* Wilkinson, D.G., Ed., IRL Press, Oxford, 1992, 75.
33. Karnovsky, M.J., A formaldehyde-glutaraldehyde fixative of high osmolarity for use in electron microscopy, *J. Cell Biol.,* 27, 137A, 1965.
34. Sulik, K.K., Johnston, M.C., Daft, P.A., Russell, W.E., and Dehart, D.B., Fetal alcohol syndrome and DiGeorge anomaly. Critical ethanol exposure periods for craniofacial malformations as illustrated in an animal model, *Am. J. Med. Genet.,* Suppl. 2, 92, 1986.
35. Augustine, K.A., Liu, E.T., and Sadler, T.W., Interactions of *Wnt-1* and *Wnt-3a* are essential for neural tube patterning, *Teratology,* 51, 107, 1995.
36. Augustine, K.A., Liu, E.T., and Sadler, T.W., Antisense inhibition of *Engrailed* genes in mouse embryos reveals roles for these genes in craniofacial and neural tube development, *Teratology,* 51, 300, 1995.
37. Sadler, T.W., Liu, E.T., and Augustine, K.A., Antisense targeting of *Engrailed-1* causes abnormal axis formation in mouse embryos, *Teratology,* 51, 292, 1995.
38. Satokata, I. and Maas, R., *Msx-1* deficient mice exhibit cleft palate and abnormalities of craniofacial and tooth development, *Nat. Genet.,* 6, 348, 1994.
39. Davidson, D.R., and Hill, R.E., *Msh*-like genes: a family of homeobox genes with wide-ranging expression during vertebrate development, *Semin. Dev. Biol.,* 2, 405, 1991.
40. Robert, B., Sassoon, D., Jacq, B., Gehring, W., and Buckingham, M., *Hox-7,* a mouse homeobox gene with a novel pattern of expression during embryogenesis, *EMBO J.,* 8, 91, 1989.
41. Ang, S.L. and Rossant, J., *HNF-3b* is essential for node and notochord formation in mouse development, *Cell,* 78, 561, 1994.
42. Weinstein, D.C., Ruiz i Altaba, A., Chen, W.S., Hoodless, P., Prezioso, V.R., Jessell, T.M., and Darnell, J.E., Jr., The winged-helix transcription factor *HNF-3b* is required for notochord development in the mouse embryo, *Cell,* 78, 575, 1994.
43. Sasaki, H. and Hogan, B.L.M., Differential expression of multiple fork head related genes during gastrulation and axial pattern formation in the mouse embryo, *Development,* 118, 47, 1993.
44. Echelard, Y., Epstein, D.J., St-Jacques, B., Shen, L., Mohler, J,. McMahon, J.A., and McMahon, A.P., Sonic hedgehog, a member of a family of putative signaling molecules, is implicated in the regulation of CNS polarity, *Cell,* 75, 1417, 1993.
45. McMahon, A.P., Joyner, A.L., Bradley, A., and McMahon, J.A., The midbrain-hindbrain phenotype of *Wnt-1/Wnt-1* mice results from stepwise deletion of *engrailed*-expressing cells by 9.5 days postcoitum, *Cell,* 69, 581, 1992.
46. Thomas, K.R. and Capecchi, M.R., Targeted disruption of the murine *int-1* proto-oncogene resulting in severe abnormalities in midbrain and cerebellar development, *Nature,* 346, 847, 1990.
47. Foerst Potts, L. and Sadler, T.W., The role of *Msx-1* and *Msx-2* in development using antisense technology with murine whole embryo culture (abstract), *Teratology,* 51, 156, 1995.

Appendix: Supplies and Suppliers

Whole-Embryo Culture

- Rotator wheels (Scientific Industries; Bohemia, NY) consisting of all-purpose head assembly (one head assembly will hold 15 culture bottles), 11- to 18-mm clips
- Rotator motors, Model SZ809, 30 rpm (Dayton Manufacturing Electric Co.; Chicago, IL)
- Dayton pillow bloc, ball bearing bore flange type (Dayton Manufacturing Electric Co.; Chicago, IL)
- Convection-type incubator, minimum door opening 18 × 18 inches
- Dissecting microscope, base illuminated
- Serum vials, 10 and 30 ml (Fisher Scientific; Raleigh, NC)
- Silicone stoppers, size 0 (VWR Scientific; Atlanta, GA)
- Forceps, Dumont number 5 (Biomedical Research Instruments, Inc.; Rockville, MD)
- Dissecting scissors
- Compressed gas tanks: 5% CO_2, 5% O_2, 90% N_2; 5% CO_2, 20% O_2, 75% N_2; 5% CO_2, 95% O_2 (v/v)

Microinjection

- 10 μl VWR Microdispenser 100 replacement tubes (VMR Scientific; San Francisco, CA)
- PN-3 Glass Microelectrode Puller (Narisnige Scientific Instrument Lab; Tokyo, Japan)
- MF-83 Microforge (Narisnige Scientific Instrument Lab, Tokyo, Japan)
- Injection apparatus consisting of Yale hypodermic 10-cc syringe with Lver Lok tip (Becton Dickinson; Franklin Lakes, NJ) attached to a 19-gauge needle connected to thin silastic tubing

PCR

- DNA thermal cycler (Perkin Elmer; Branchburg, NJ)
- Gene Amp RNA PCR core kit (Perkin Elmer; Branchburg, NJ)
- Electrophoresis gel apparatus

Whole-Mount In Situ *Hybridization and Immunohistochemistry*

- Multiwell Tissue Culture 24-Well Plate Falcon 3047 (Becton Dickinson Labware; Lincoln Park, NJ)
- S/P Rotator V (Baxter; Charlotte, NC)

- Maxi Hybridization Oven (Labnet; Woodbridge, NJ)
- Genius 3 Nonradioactive Nucleic Acid Detection Kit and Genius 4 Nonradioactive RNA Labeling Kit (Boehringer Mannheim Biochemicals; Indianapolis, IN)

Histology and SEM

- Araldite 502 Resin and Propylene Oxide (Ted Pella, Inc.; Tustin, CA)
- 7800 LKB Bromma knifemaker
- 2088 Ultrotome V, LKB Bromma
- CPD 010 Balzers Union FL-9496 Balzers critical point dryer
- Gold SEM Coating Unit EF100
- JEOL 35B SEM

Chapter

Nucleic Acid Amplification Technologies

Richard H. Finnell, Scott J. Vacha, and Scott A. Mackler

Contents

0-8493-3342-3/97/$0.00+$.50

I. Introduction

Long before the association was made between *in utero* exposure to thalidomide and an increased incidence of congenital limb reduction malformations,[1-3] teratologists and developmental toxicologists have been seeking mechanistic explanations for the action of various teratogenic compounds on embryonic development. Although these studies have been largely limited to morphological and, more recently, to biochemical investigations, these studies have provided a great deal of what we know today about abnormal morphogenesis. In the last few years, advances in molecular biological techniques have made these state-of-the-art experimental approaches increasingly available to an ever widening body of scientists, including those interested in learning more about the adverse effects of environmental agents and pharmaceutical compounds on developing fetuses.

The use of modern molecular biological tools by teratologists has created the situation where interactions between teratogenic agents and embryonic gene targets which had been long sought after could finally be addressed. When teratogens were observed to induce specific major malformations, such as cortisone-induced cleft palate,[4] acetazolamide-induced ectrodactyly,[5-6] or valproic acid-induced neural tube defects,[7-8] there was a natural inclination to consider the possibility that the expression of selected embryonic morphoregulatory genes was being adversely affected by the teratogenic treatment. Furthermore, using the newer technologies has made it possible to determine which cell types are particularly susceptible to the teratogenic insult and whether the effect was anatomically restricted or generalized throughout the embryo. Once it can be established that the teratogenic insult alters gene expression, then it is possible to identify which of the many potentially important genes are being affected. It also may be possible to determine if the altered gene expression serves to explain strain differences in response to a teratogenic treatment which are occasionally observed in teratological investigations. From that point, it becomes possible to ultimately determine how the observed changes in gene expression results in abnormal morphogenesis.

The key to approaching any of the above-mentioned questions by developmental toxicologists rests on the ability to have sufficient material with which to work. That is, such experimentation requires a ready source of RNA and DNA. In the past, obtaining sufficient quantities of such molecules, especially from embryonic material, was extremely problematic. However, there have been significant advances in nucleic acid amplification technologies that now make it possible to obtain working quantities of nucleic acids from starting materials as limited as a single cell. These techniques include the well known polymerase chain reaction (or PCR) and *in situ* transcription coupled with an anti-sense RNA amplification procedure.[9-11] This latter approach is an emerging technology that has been widely applied in the neurosciences and is only now being more frequently utilized by developmental biologists. This chapter will review these nucleic acid amplification approaches, providing not only the theory behind how they work, but also detailed explanations of various experimental applications. Complete experimental protocols have been appended to the end of the chapter.

II. The Polymerase Chain Reaction

The polymerase chain reaction is a powerful amplification technique that has revolutionized the study of gene expression in both normal and abnormal tissues. The isolation and cloning of thermostable DNA polymerases from organisms that survive in environments with high (>70°C) temperatures have permitted many laboratories to add PCR amplification to their studies (Table 5.1).[12] PCR amplification is, in theory, a simple process; however, in practice it can be complicated by many technical limitations and errors. This section will discuss the essential information that is necessary to perform PCR amplification, review many of the important technical factors that have arisen in the past decade, and illustrate its potential impact when applied to teratological studies. Several excellent articles discuss PCR amplification and its applications in more detail.[13-19]

TABLE 5.1
Common Applications of PCR[a]

Cloning of cDNAs (based on sequence similarities or differential expression; use of tissues or single cells; as an aid in clinical and forensic medicine)
Cloning of full-length cDNAs from partial-length cDNAs (anchored PCR, *RACE,* and modifications)
Direct DNA sequencing
Synthesis of labeled probes for hybridization studies (e.g., library screen) and primer extension
Site-directed and random mutagenesis
Library construction
Quantitative measurement of low abundance mRNAs
Demonstration of genetic mutations (e.g., *SSCP*)
In situ PCR

[a] Other applications for PCR exist and undoubtedly more will be developed in the future

A. Theoretical Considerations

The three steps of conventional PCR amplification are (1) denaturation of the template DNAs, (2) annealing of the oligonucleotide primers, and (3) cDNA synthesis. Separation of double-stranded DNA into single strands is necessary for hybridization of the primers, and this is performed at a high temperature (>90°C). The primers are allowed to hybridize to these single-strand DNAs in the next step. The selected temperature is a function of the base composition of the oligonucleotides and their complementarity to the DNA templates. Polymerization occurs in the third step, at a temperature near the optimal temperature for the thermostable enzyme but not so high that the primers will "fall off" before initiation of cDNA synthesis. One set of these three sets is called a cycle. The first three cycles of PCR amplification from a single cDNA template are shown in Figure 5.1. The power of PCR amplification is that it proceeds geometrically. If all steps occur properly and the efficiency is 100%, then the amplification will result in a 2^n-fold increase in the amount of the original template (n = the number of cycles); 25 cycles will amplify the original molecule 2^{25}, or greater than 30-millionfold.

Requirements for PCR amplification consist of a template DNA that contains the target sequences, oligonucleotide primers, the four different dNTPs, a thermostable DNA polymerase, a buffer, KCl, $MgCl_2$, and a thermal cycler machine. Either cDNA synthesized from RNA (known as reverse transcription PCR, or just RT-PCR) or genomic DNA may be used as the template for amplification. The template DNA should be free of any potential contaminants and, if made from RNA, be of sufficient length to include the region to which the primers are directed. The primers will determine the limits of the cDNA products (Figure 5.1), and they should be designed so as to not form hybrids with themselves or each other. If the goal is to amplify a known mRNA, then selecting primers that are on opposite sides of an intron will ensure that the cDNA product was not amplified from genomic DNA (in this example PCR cDNAs made from genomic DNA will be larger than the predicted size; see Figure 5.2). The primers and dNTPs must be present in excess amounts so as not to limit cDNA synthesis; however, dNTPs present in vast excess can lead to the formation of nonspecific PCR products. The various properties of thermostable DNA polymerases are listed in Table 5.2, and more than six polymerases are currently available. The optimal Mg^{++} concentration must be determined for each specific PCR, and this value usually ranges from 1.0 to 2.5 m*M*. Many different types of thermal cyclers are currently available. The majority of cyclers work by heat conduction, although convection ovens are also commonly used. Some of the newer thermal cyclers exploit Pelletier physics to improve the speed of the heat exchange. Important considerations in selecting a thermal cycler include: cost, speed (the time to complete 30 typical cycles can range from 60 minutes to more than 6 hours, depending on the type of machine selected for use), and the total number of sample tubes that can be included for each run. *In situ* PCR requires certain adapters or modifications for the thermal cycler.

B. Sample PCR Protocol

Equipment used for PCR should be kept in closed containers and away from possible contamination by exogenous DNA (found on the skin, in test tubes and other items, or in aerosolized particles). The pipettors should be dedicated for PCR use only and, if possible, work by positive pressure displacement.

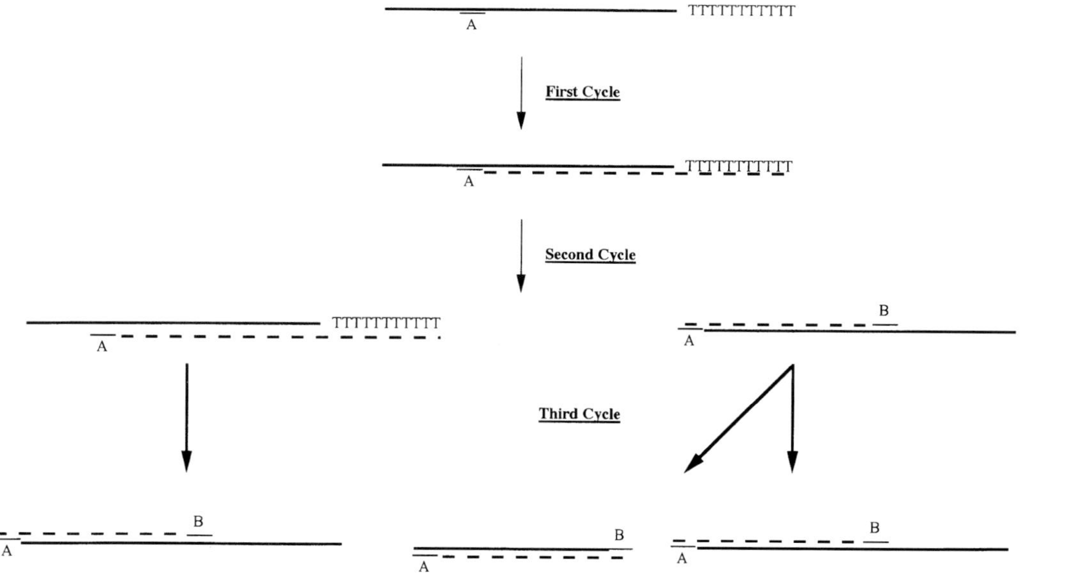

FIGURE 5.1
PCR Amplification. An example is shown of a single cDNA molecule using primers that are labeled A and B. The cDNA was made by reverse transcription from a single mRNA molecule and the oligothymidine region of the cDNA is its 5′ end. In the first cycle, A primes synthesis of a cDNA molecule (dotted line) that can extend to the end of the template (a "long" product). Primer B initiates synthesis of a cDNA on this "long" product in the next cycle and extends to a region complementary to primer A (a "short" product). Another "long" product is synthesized from the original cDNA template in this same cycle. Completion of the third cycle results in eight molecules, including a double-stranded "short" cDNA. This is the expected PCR product and amplification will now proceed geometrically. Synthesis of another "long" product with the original template and primer A is not shown in the third cycle, although it does occur. The "long" cDNAs are amplified linearly and will constitute a very small percentage of the total of amplified cDNAs. The dotted lines represent the polymerase step of a cycle, and this always proceeds in the 5′ to 3′ direction.

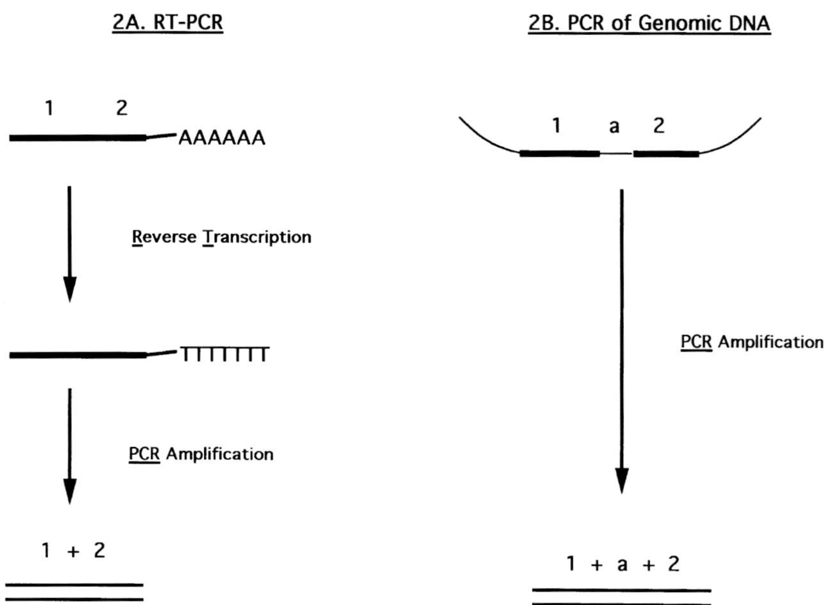

FIGURE 5.2
PCR Amplification of a cDNA from mRNA or genomic DNA that contains an intron. **(A)** Reverse transcription (RT) of mRNA with oligothymidine primer and amplification using primers that hybridize to two different exons (1 and 2) results in a cDNA product of a predicted size. This cDNA is shorter than that synthesized using the same primers and genomic DNA containing an intron (a) between the exons. **(B)** Design of the primers on opposite sides of an intron, when possible, helps to ensure that genomic DNA did not contaminate the sample if the goal is to amplify cDNA from mRNA.

TABLE 5.2
Properties of Thermostable DNA Polymerases

Misincorporation rate ("fidelity" or proof-reading via 3′-5′ exonuclease activity)
Length of cDNAs to be synthesized/processivity rates
Formation of blunt-end or adenosine "overhang" (important for subcloning; 3′-5′ exonuclease activity will result in mostly blunt-end products)
Stability at high temperatures (half-life at >90°C)

Pipette tips with filters that prevent aerosol particles from entering the pipette chamber are helpful and can be used with negative-displacement pipettors. Careful laboratory methods to minimize "carryover" contamination with DNA are critical for the success of any PCR experiment. The preparative benchwork for PCR amplification may be performed in three steps: reaction preparation, thermal cycling, and analysis of results.

1. Reaction Preparation

Reagents should be frozen in aliquots for use in a limited numbers of reactions. The stock solutions include dATP, dGTP, dCTP, and dTTP (10 m*M* is a convenient concentration, and they may be frozen separately or together); Mg^{++}-free buffer for the thermostable polymerase, and $MgCl_2$ (25 m*M*). Aliquots of the buffer are provided by the manufacturer with each thermostable DNA polymerase and typically include a Tris buffer, KCl, and bovine serum albumin. Take care to note that the buffer is Mg^{++}-free. Specific oligonucleotide primers are used in excess in PCR (to avoid limiting cDNA synthesis) and the final concentration should be >0.1 μ*M*. The oligonucleotides and DNA templates may be kept in TE buffer; however, the EDTA concentration should not be high in PCR (>20 μ*M*) or significant chelation of Mg^{++} may occur.

Procedure

1. Add to each sterile tube (0.5 to 1.5 ml size) everything except the DNA template. Final concentrations include the buffer (1×), dNTPs (at least 5 μ*M* of each, but higher concentrations may be used) and oligonucleotide primers (>0.1 μ*M*). A master mix may be prepared for each set of reactions and aliquoted into separate tubes. Volumes of a typical reaction can range from 25 to 100 μl.
2. Determine the optimal Mg^{++} concentration for each set of oligonucleotide primers. Select a range (for example, from 1.0 to 3.0 m*M*) and perform reactions in separate tubes with only the Mg^{++} concentration altered.
3. Always include a tube with everything necessary for PCR amplification except for a DNA template. This is used as a negative control to ensure that exogenous DNA has not contaminated the preparation. Analysis of the contents of this tube will show the oligonucleotide primers, but no cDNAs should be present.
4. The template DNA (usually contained within a cDNA synthesis reaction or in a sample of genomic DNA) is added last to the reaction mixture. Many thermal cyclers require a layer of a small amount of mineral oil over the reaction volume in order to prevent evaporation during the multiple steps at high temperatures. The mineral oil does not significantly interfere with the reaction, but it can be messy. Make sure that the stock solution of oil does not become contaminated with exogenous DNA. Thermocyclers with an enclosed heated cover are one example of machines that do not require mineral oil in the individual tubes.

2. Thermal Cycling

Procedure

1. Program the thermal cycler for the desired temperature and duration of the denaturation, hybridization, and polymerization steps. Each machine has instructions on how to program the heating and/or cooling elements. In most cases, it is helpful to initially denature the template DNA for a longer period (5 minutes at 95°C) and cool quickly on ice before adding to the reaction mixture. The

denaturation step of the complete cycle should be at 90 to 95°C, and no more than 60 seconds are necessary (long periods at high temperatures eventually will decrease the activity of the DNA polymerase). The temperature for annealing of the primers is a function of the calculated T_m (the temperature at which half of the oligonucleotides are annealed to the DNA and half are free in the solution). An estimate of the T_m for each oligo equals the sum of the following equation: 4 times the number of G and C residues, plus 2 times the number of A and T residues. First try an annealing step that is close to the T_m-25°C. Often it is necessary to either increase or decrease the annealing temperature, depending on the specific results. The duration of the annealing step can range from 30 to 90 seconds and also may additionally require testing of different intervals. The polymerization step is performed at the optimal temperature for the enzyme (typically 72°C) and should be long enough in duration to allow for complete cDNA synthesis. Single nucleotides are incorporated as quickly as 2 to 4 Kb per minute, and 60 to 90 seconds is often sufficient time for DNA synthesis. These three steps (denaturation, annealing, polymerization) comprise one complete cycle. Most PCR amplifications need no more than 30 to 35 cycles. If another repetition of cycles is to be performed, add another small amount of the DNA polymerase. Most investigators finish with an extended time at 72°C to ensure that the final cycle cDNA products are synthesized in their entirety (theoretically 50% of the total cDNA molecules).

2. Select and run the program. A cautious person intermittently checks the temperature display to ensure that the program has been properly set and is being executed by the thermocycler. The tubes should be placed on ice or frozen at the end of the program, although letting the tubes sit overnight at room temperature should not pose major problems.

3. Specific Technical Considerations

PCR amplifications often result in indeterminate results or multiple cDNA products. Some of these cDNAs are artifactual or represent false-positives due to exogenous DNA contamination. Many factors can contribute to these problems (Table 5.3), and the careful person is able to identify the source of a problem with minimal effort. The optimal magnesium concentration, as discussed earlier, should be determined with each set of experiments. Excess dNTPs can result in nonspecific products. Their concentration should be sufficient so as not to limit synthesis, but should not be in such abundance that it contributes to nonspecific cDNAs.

Selection of the optimal temperature for hybridization of the primers can be problematic and require a trial and error approach. Although the estimated T_m is used initially, many different temperatures may need to be examined. This can be quite time consuming, as it requires a new round of PCR amplification for each attempt. Primer oligomerization and/or mispriming of the DNA template may occur before the initial denaturation. Efforts to limit this problem include separation of some components before the first polymerization. These include a solid barrier with wax that melts in the first step or a monoclonal antibody directed against the thermostable polymerase which is destroyed during the initial denaturation.

TABLE 5.3
Important Technical Aspects of PCR

Mg^{++} concentration
dNTP concentration
Quality of template DNA
Design and amount of primers
Temperature selected for annealing of primers to template
Initial cycle conditions (to increase specificity of amplification; use of wax beads or monoclonal anti-Taq antibody)
Choices of thermostable enzymes
Prevention of DNA contamination ("false positive" results)

Perhaps the major technical problem is the synthesis of false positives. As a result, cDNAs are amplified because of contamination of the sample with exogenous DNA. The tremendous power of PCR means that single molecules inadvertently introduced into the reaction tube may be detected with the chosen oligonucleotides. A negative control should be included in every amplification. Several physical measures may be employed to avoid this contamination, including separate work areas, dedicated pipette equipment, and ultraviolet lighting, to name but a few.

4. Interpretation of the Results

A small amount (typically less than 10% of the volume) of the PCR is analyzed for the presence of cDNAs by electrophoresis in an agarose gel. In most instances the cDNAs can be visualized under ultraviolet illumination after staining with ethidium bromide. Smaller cDNAs (less than 150 base-pairs) may require size separation in a polyacrylamide gel. In certain cases, the PCR products may be present in picogram or less amounts and only detectable after incorporation of a radioisotope and exposure to autoradiographic film (e.g., differential display PCR).

The presence of any cDNA after PCR does not always ensure that the desired product was synthesized, even when the cDNA is the size predicted by the choice of oligonucleotide primers. Many other cDNAs can be synthesized, and it is critical to demonstrate the identity of any of the PCR cDNAs that may be of importance to the experiments. This can be accomplished by several methods, each of which requires some knowledge of the DNA sequence. These methods include hybridization of a nucleic acid probe, use of specific restriction endonucleases and sequencing of the cDNA.

Hybridization uses a probe that is complementary to the region between the 3′ and 5′ ends defined by the PCR primers. Hybridization of this probe to a Southern blot of the agarose gel that contains the PCR cDNA under appropriate stringency conditions indicates that the PCR product contains a nucleotide sequence that was synthesized during PCR. In some cases, specific restriction endonucleases sequences are present inside the 3′ and 5′ ends of the PCR product. Successful digestion of the PCR product into predicted sizes also indicates that PCR amplification synthesized the desired cDNA. The most direct method of determining the identity of the PCR cDNAs is DNA sequencing, but this can be highly time and labor consuming. Available choices are sequencing the cDNA directly during or after PCR (in this case the PCR should result in one specific cDNA product) or subcloning the cDNAs into a vector followed by DNA sequencing. It is critical, no matter the method used, to determine the identity of the cDNA(s) after PCR amplification.

III. Differential Display

In addition to previously discussed applications, the polymerase chain reaction has direct applications in the area of differential gene expression. The use of PCR for this purpose was originally developed and coined "differential display" in 1992.[17] Because the ability to identify genes with altered expression patterns is important to understanding normal embryogenesis in many experimental systems, differential display has become a valuable research tool.

Of the entire mammalian genome, it is estimated that only 15% is expressed at any given time.[20] While a seemingly small percentage, this amounts to approximately 10,000 to 20,000 different mRNAs in a given mammalian cell.[21] Differential display was designed to provide a visual representation of these mRNAs, just as two-dimensional gel electrophoresis has been used to visualize protein expression patterns. By comparing the resulting autoradiographic patterns of mRNAs across treatment or comparison groups, differentially expressed genes can be detected and cloned for further characterization. The power of the differential display approach is that it allows for the screening of literally thousands of genes, with the potential to identify novel genes that may be transcriptionally regulated in any given experimental system. As with any new technique, differential display offers a myriad of potential applications. Studies of interest to the developmental toxicologist include teratogen-induced gene alterations, regional specific gene patterns during development, or the identification of sensitivity/resistance genes to different teratogen-induced malformations.

Other techniques have been available to address differential gene expression,[22-25] including differential hybridization and subtractive hybridization strategies. However, differential display PCR offers a major advantage over these approaches. The reactions may be performed with very little mRNA as the starting material because of the power of PCR amplification. This enables the investigator to use specific targets, including developing embryonic structures and even single cells.[26] In addition, differential display PCR is a relatively quick procedure and cDNAs can be subcloned for further characterization within two weeks.

Differential display involves amplifying an entire mRNA population, followed by electrophoresis through denaturing acrylamide gel for pattern comparison

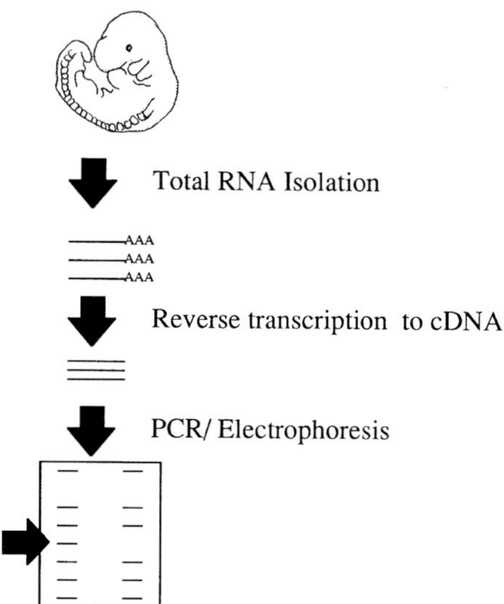

FIGURE 5.3
Overview of differential display. Total RNA is isolated from an embryonic tissue source, reverse transcribed to cDNA, and amplified by PCR. Differentially expressed cDNAs (arrow) can be visually resolved after PAGE and isolated for further characterization.

(Figure 5.3). From this simple explanation, two technical difficulties are readily apparent. Typically, PCR is used to amplify a single nucleic acid fragment of known size and sequence. However, in this approach, the amplification includes a population of mRNAs of differing sizes and of unknown sequences. As a result, specific primers must be designed to amplify the entire mRNA pool. The second difficulty involves the interpretation of amplified fragments. It is imperative that the primers be selected so that only a subset of the mRNA population is amplified, so that distinct bands can be visualized and pattern comparisons can be made across experimental groups.

In order to perform differential display, it is necessary to have at least 250 ng of total RNA from the source tissue of interest. While many procedures are available for RNA isolation,[27-29] the guanidinium thiocyanate method is probably the most common. Once the RNA has been isolated, it is divided into four microcentrifuge tubes for reverse transcription to cDNA. This is performed using AMV reverse transcriptase (Seikagaiku; Bethesda, MD) and a specific $T_{12}MN$ primer. As the abbreviation suggests, the $T_{12}MN$ primer is an oligonucleotide with a specific 3′ sequence; M is a degenerate base (adenine, cytosine, or guanine), and N refers to any one of the four nucleotides. The poly $d(T)_{12}$ portion of the primer will anneal to the poly A tail of mRNAs, while the 3′ base of the oligomer serves to anchor it specifically to a subset of mRNAs containing a complementary 3′ sequence. Consequently, by using four separate primers ($T_{12}MA$, $T_{12}MC$, $T_{12}MG$, $T_{12}MT$), the entire mRNA population is theoretically transcribed as four different subpopulations of cDNA. The penultimate base of the 14mer is not a factor in the oligomer's specificity, as it is considered a wobble base.[30]

Once the mRNA pool has been converted to cDNA it is ready for amplification by PCR. PCR amplification utilizes two primers, the same $T_{12}MN$ primer used in reverse transcription on one strand and a second 10mer as the opposing primer (Figure 5.4). By amplifying with a radiolabeled dNTP and several different primer combinations (Figure 5.5), a percentage of the mRNA population can be profiled. Although increasing primer combinations will allow the amplification of additional subpopulations of the original RNA, redundant products will appear due to the potential for different 10mers to amplify the same cDNAs. The resulting PCR products are then run on a 6% denaturing acrylamide gel, such that samples of identical primer combinations are loaded in adjacent lanes for comparison. Following electrophoresis, the differential display gel is exposed to X-ray film to resolve the pattern of message expression.

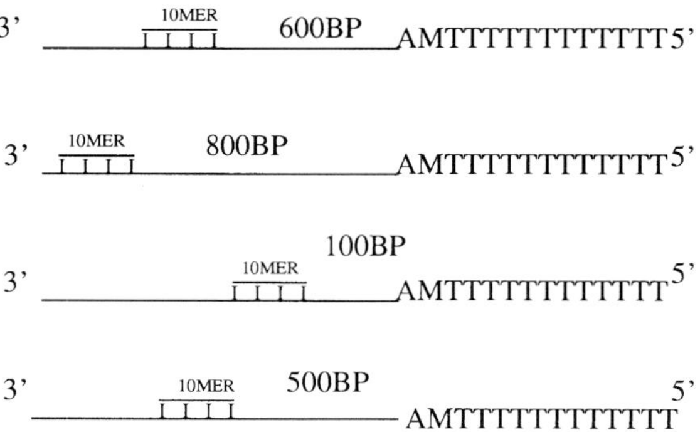

FIGURE 5.4
The use of arbitrary 10mers for differential display PCR results in products of differing sizes. This is due to the random annealing of the primer and the low stringency annealing temperature used in the PCR.

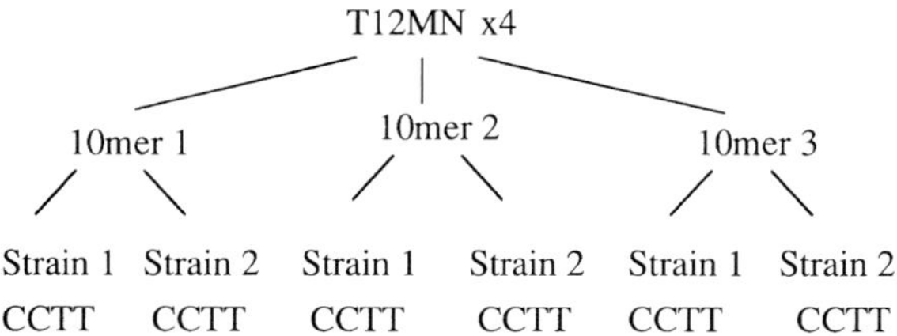

FIGURE 5.5
Sample differential display PCR primer combinations comparing two inbred mouse strains exposed to two teratogenic treatment conditions (C = 38°C control treatment; T = 43°C hyperthermia treatment).

Figure 5.6 illustrates a representative differential display gel. This gel compares two highly inbred mouse strains (SWV, LM/Bc) differing in their susceptibility to hyperthermia-induced neural tube defects.[31] Duplicate lanes of the embryonic RNA (cDNA on the gel) have been run following *in utero* exposure to either a 38°C control or a 43°C hyperthermia treatment. The bands present in all lanes (Figure 5.6C) are considered to be genes which are expressed in all treatment groups, while bands present in individual lanes are considered to be uniquely expressed (Figure 5.6A and D). For example, the arrow for (D) in Figure 5.6 indicates a mRNA species that is uniquely expressed in the LM/Bc strain; however, its expression is not affected by heat treatment. On the other hand, the arrow for (A) in the same figure indicates a band the expression of which is unique to one strain and changes with teratogenic treatment. Using differential display PCR in this manner, it may be possible to identify genes conferring either resistance or susceptibility to different teratogens.

Once a gene has been identified as being differentially expressed, it must be isolated from the acrylamide gel for further characterization (Figure 5.7). This is accomplished by overlaying the developed film onto the acrylamide gel and simply removing the associated band with a razor blade. The gel and film must be aligned properly so as to avoid accidentally removing additional bands adjacent to the band of interest.[32] This can be accomplished by hole punching the gel and film prior to exposure or by the use of radioactive ink markings.

While obtaining unique bands utilizing this experimental technique is relatively simple, the verification and further characterization of isolated bands is extremely labor intensive. The high number of false positives attributed to differential display are thought to be due primarily to chromosomal contamination[30] or a heterogeneous RNA source.[33] It should be noted that because a heterogenous tissue source may result in banding variability (Figure 5.6B), running duplicate lanes may be helpful in resolving the number of false positive bands that appear on these gels. Confirmation of isolated genes generally is obtained by Northern blot analysis[34] against the original RNA source, while further characterization may involve cloning, DNA sequencing,[35] *in situ* hybridization,[36] and other experimental techniques depending on the research context.

Since the initial development of differential display, several improvements and procedural alterations have been made which deserve attention. The primary improvements have involved the selection of primers.[37] New primer strategies have included the elongation of the arbitrary primers and addition of restriction sites for greater stringency and cloning capability,[38] end-labeling the primers for greater sensitivity,[39] the use of one-base anchored primers,[40] and the use of extended primer sequences to target genes with specific binding domains.[41] There has also been some discussion regarding the type of radioisotope used. While [^{35}S]dATP was originally used by Liang and Pardee, [^{33}P]dATP has been suggested as the isotope of choice. [^{33}P]dATP shows

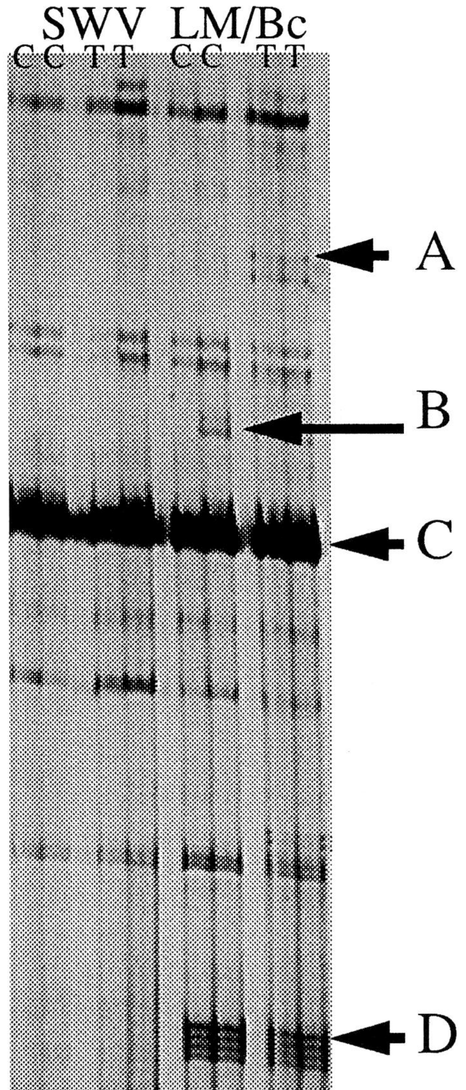

FIGURE 5.6
Representative differential display PCR gel showing embryonic strain differences following *in utero* exposure to either a 38°C control treatment or a 43°C hyperthermia treatment. Arrows A through D are explained in the text.

similar sensitivity without the resulting volatile decomposition product observed with [^{35}S]dATP.[42] Other improvements include additional means for direct sequencing of PCR products[43] and slot blot screening of differential display fragments.[44]

While the differential display methodology is continually being refined, it has already proven to be a highly valuable research tool. Within three years

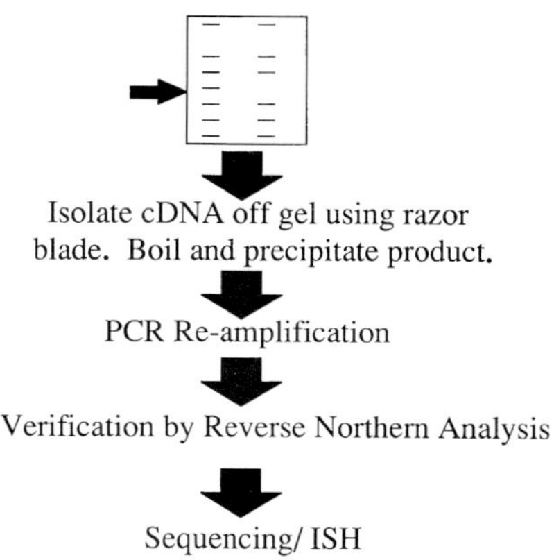

FIGURE 5.7
Isolation and characterization of differentially expressed cDNAs. Bands are cut from the acryalmide gel and reamplified for cloning, Northern blotting, and sequencing.

of its development, it has been used successfully to identify genes involved in cancer[17, 30, 45-47] and tumor development,[48] growth factor induction,[41, 49] early embryogenesis,[50] and other treatment-induced transcriptional studies.[51-56] Most recently, it has been used in cultured rat fetal hippocampal cells to provide direct evidence that a large number of mRNAs, including those of the glutamate receptor family and second messenger systems, are present within individual processes of the cultured cells.[26] The continued application of differential display should lead to the identification of additional genes showing transcriptional regulation in other experimental systems.

IV. *In Situ* Transcription/aRNA Amplification

Another approach to nucleic acid amplification utilizes both *in situ* transcription and anti-sense RNA amplification in order to obtain abundant yields of desired targets. *In situ* transcription is a powerful technique that synthesizes cDNA from anatomically-restricted regions; therefore, it is ideal when used as the first step in the creation of enriched cDNA libraries from specific regions of the developing embryo. The *in situ* transcription procedure involves the annealing of a specific amplification oligonucleotide primer to the polyadenylated mRNA population within the starting embryonic material. cDNA synthesis is initiated by the addition of reverse transcriptase and deoxynucleotide triphosphates. We have utilized *in situ* transcription together with aRNA amplification procedures to examine the impact of teratogenic treatment on the coordinate expression of selected genes that are important to the embryo's ability to complete normal neural tube closure and craniofacial development. This portion of the chapter is devoted to the necessary procedural considerations of this type of experimentation. Actual experimental protocols are provided in the appendix at the end of this chapter.

To examine changes in embryonic gene expression secondary to a maternal teratogenic exposure, one need only collect embryonic or fetal material of interest following the desired teratogenic insult. When pursuing such investigations, it is important to consider the kinetics of the teratogen in question and the interval between the treatment and collection of the embryos. Certain mRNA species, such as the stress protein mRNAs, may be induced rapidly and then return to basal levels relatively quickly, which would necessitate embryo harvest within a shorter interval than for other potential genes of interest.[57] The collected material that is to be used as a tissue source could be the whole embryo or an isolated embryonic structure, such as the neural tube, a branchial arch, a palatal shelf, or even a limb bud. This experimental procedure has a great deal of flexibility, so that it is up to the investigator whether or not to fix the embryos and cut sections (up to 8 μm) from paraffin-embedded specimens on a microtome which then can be mounted on gelatin subbed slides[11] or to quick freeze the embryos in liquid nitrogen for subsequent cryostat sectioning. The advantage of using material fixed to slides is that it is possible to identify the specific anatomical structure of interest and eliminate potentially contaminating tissue types by physically removing the undesired tissue from the slide. It is also possible to get a visual representation as to where the gene of interest is being expressed by exposing the slides to X-ray film if synthesis incorporated a radiolabel and to have material available for subsequent characterization by either *in situ* hybridization or immunohistochemistry. From a standpoint of experimental practicality, the ability to make use of fixed specimens is extremely important, for not only is it possible to collect specimens and store them for use at a more convenient time, but it is also possible to make excellent use of existing stored specimens that have been collected for any number of reasons over the years and find new and exciting experimental questions to ask without the need to use additional experimental animals for the collection of new specimens.

The use of fresh nonfixed or frozen embryos is yet another possibility, as one can immediately isolate the desired tissues manually, using sharpened tungsten needles for the microdissection. Although it is not absolutely necessary, it sometimes helps to use a gentle treatment with collagenase prior to the microdissection.[58] The tissue can be collected directly into an microcentrifuge tube and the cells sonicated and permeabilized by the addition of digitonin.[59] A small amount of RNasin® is added by sonication to the hybridization buffer to inhibit the degradation of the RNA by endogenous Rnases. Once the desired embryonic material has been obtained and isolated from contaminating sources of mRNAs, the tissue is primed with an unlabeled oligo-(dT_{24})-T7 oligonucleotide [AAACGACGGCCAGTGAATTGTAATACGACTCACTATAGGCGC(T_{24})] containing the promoter region for T7 RNA polymerase (Figure 5.8). This reaction is allowed to proceed for 90 to 180 minutes at 37°C. The amplification oligonucleotide primer will hybridize to the tissue's poly A^+ mRNA, which represents greater than 98% of all of the functionally active genes. The hybridization is followed by the addition of avian myeloblastasis reverse transcriptase (Seikagaiku; Bethesda, MD) and

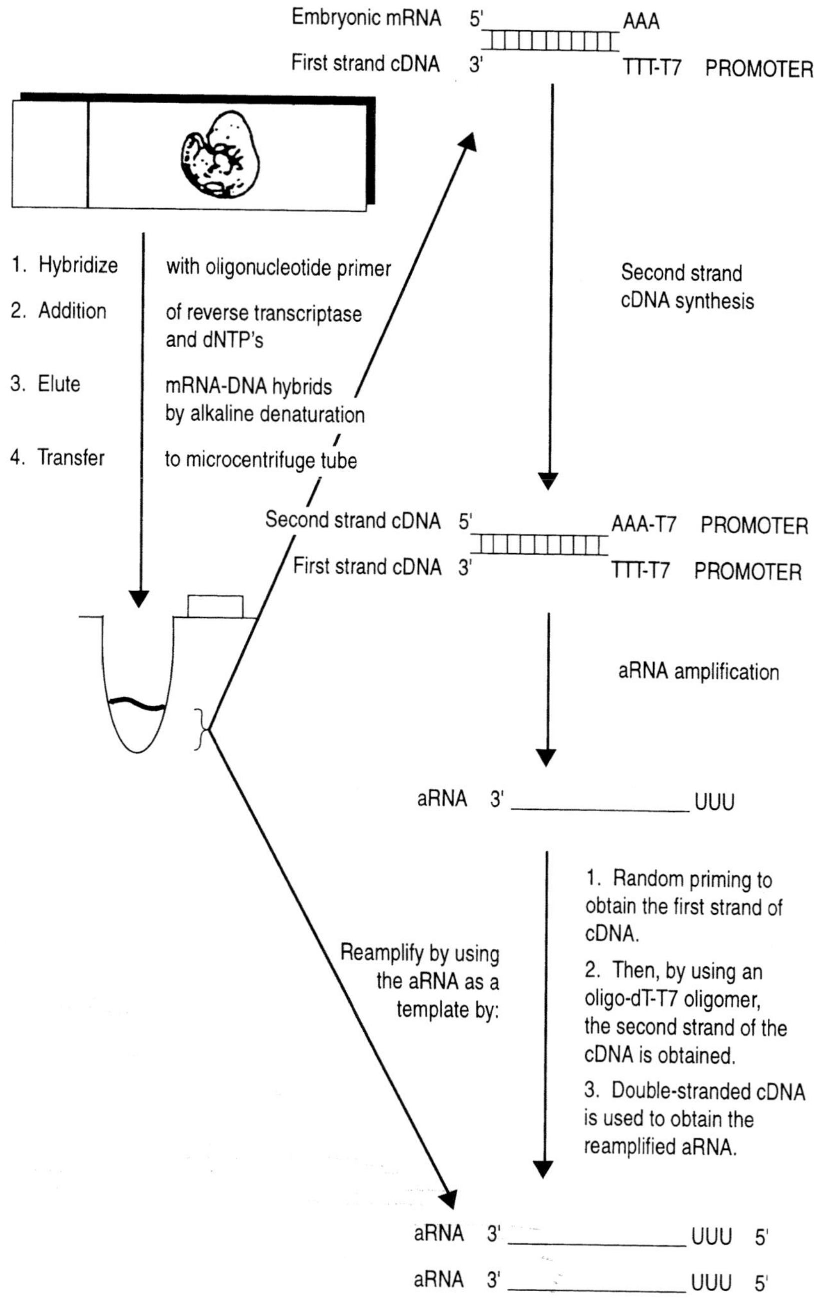

FIGURE 5.8
Schematic representation of the *in situ* transcription and amplified antisense RNA amplification procedure from a fixed and sectioned embryonic tissue source.

both labeled and unlabeled deoxynucleotides. This reaction copies endogenous poly A^+ mRNA into cDNA along with the T7 promoter, while preserving the anatomical distribution of the cDNA transcripts. This allows for future studies to be conducted within specific regions of any structure in the developing embryo.

The *in situ* transcribed cDNAs can be removed from the sections by the alkaline denaturation of the mRNA-cDNA hybrids, whereas when using isolated fresh tissue, this alkaline denaturation will free the cDNA to become double-stranded. This involves the addition of 0.5 *N* NaOH and 0.1% (w/v) sodium dodecyl sulfate and agitating the tissue. A small volume of 1 *M* Tris (pH 7.0) and 1 µl of glycogen (4 µg/µl) as a carrier is added to the reaction mixture. The single-stranded cDNA is first recovered following a phenol:chloroform extraction and ethanol precipitation and is made double-stranded by self-priming (hairpin loop formation). This is followed by a brief, S1-nuclease treatment in order to remove the hairpin loop structure and blunt ended with T4 DNA polymerase and the large fragment of DNA polymerase I. Following the removal of unincorporated deoxynucleotides by either spin columns or drop dialysis on a Millipore (Bedford, MA) HAWP filter, the cDNA, which has an incorporated T7 promoter, is now ready for amplification using T7 RNA polymerase to drive the synthesis of amplified, antisense-RNA (Figure 5.8). Since the RNA made using the aRNA amplification technique is antisense to the original poly A^+ mRNA, it can be used directly as an RNA probe or used in other applications.[11]

The antisense RNA amplification is quite simple, involving little more than the addition of both labeled and unlabeled ribonucleotides to a limited quantity of the double-stranded cDNA template in the presence of T7 RNA polymerase (50 U per reaction; Epicentre; Madison, WI). RNasin® (Promega; Madison, WI) is also added to inhibit any RNase-induced degradation of the reaction. Each round of amplification using this approach yields approximately a 2000-fold increase in nucleic acid. That is, for each molecule of cDNA used as a template, the yield is up to 2000 molecules of aRNA after a single round of amplification. Should this yield be insufficient, it is possible to convert the aRNA back into double-stranded cDNA and then re-convert it back to aRNA by repeating the synthesis of double-stranded cDNA. This involves a phenol:chloroform extraction of the aRNA and the addition of random hexamers and reverse transcriptase for first strand cDNA synthesis. As before, the reaction incubates 90 to 180 minutes at 37°C. Following the alkaline hydrolysis to separate the single-stranded cDNA from its aRNA template, second strand synthesis proceeds with the addition of the oligo-dT-T7 amplification oligonucleotide and DNA polymerase. The newly created double-stranded cDNA is then used as a template for a second round of aRNA amplification (Figure 5.8). By going through two rounds of this cycle, typical yields may be over a millionfold from the original starting material. The size of the aRNA can be determined from a denaturing gel that will fractionate the aRNA obtained from the *in situ* transcribed embryonic material. After a single round of amplification, the aRNA typically ranges in size from a few hundred bases to over 2 kB. It has been possible to obtain some aRNAs from embryos that were in excess of 5 kB in length. Generally speaking, from our experience the aRNA produced is in excess of 90% of the length of the cDNA template from which it was generated.[11] The aRNA that has gone through two cycles of amplification tends to be somewhat shorter in length, due to the use of random hexamers to generate the new cDNA.

It is important to note that the amplified aRNA has a high degree of fidelity to the original mRNA population found in the embryo, without a great deal of skewing to give disproportionately high amounts of selected messages. The nonlinear amplification of cDNA in PCR may dramatically over-or under-represent mRNAs when compared to other mRNAs in the same tissue source. Even a small difference in the efficiencies of PCR amplification between two molecules can cause tremendous discrepencies in their relative proportions at the end of 25 to 30 cycles.[11] However, amplification of aRNA is linear, and the final products will be present in ratios that are directly proportional to those found in the original mRNA populations. Because of this, the use of PCR-generated cDNA could be problematic when used for the generation of cDNA libraries or to create probes for RNA populations. This was a very important consideration in our experiments, as the desired endpoint is an accurate depiction of the total RNA population in its original proportions as found in the embryonic tissue. It would pose a serious complication if high abundance messages were disproportionately amplified. It was also possible to check the efficiency of the amplification procedure from sample to sample by using a series of positive control cDNA constructs in known quantities, so that the linearity of the amplification of these molecules and, by analogy, the aRNA population of molecules could be determined.

These aRNA probes were used in what we refer to as "genetic expression profiling" experiments. Following IST/aRNA amplification from the embryonic substrates, the aRNA probes were allowed to hybridize to what is essentially a "reverse Northern blot". Here, the slot blot contained cDNAs encoding various genes of interest that have been immobilized to a nylon (ZetaProbe, BioRad; Hercules, CA) membrane. Identical membranes with equimolar concentrations of the cDNAs could be faithfully reproduced for probing with aRNA riboprobes. The rationale for the selection of the candidate genes that were used on the reverse Northern blots was naturally arguable. While cogent arguments could be made for a vast number of genes, it was essential to limit the selection to those genes that were appropriate to the developmental hypotheses that were being considered and that were readily available. By examining multiple blots per treatment group, over various gestational timepoints, it was possible to get a dynamic picture of transcriptional activity and detect any fluctuation in mRNA levels in response to the teratogenic exposure. Quantification of band intensity in the genetic expression profile could be performed using a variety of scanning devices, such as an optical densitometer, a phosphorimager, or a two-dimensional direct Beta detector (Ambis; Billarica, MA). The relative amounts of radiographic signal or CPM counts for each of these molecules was considered to be a direct reflection of the amount of mRNA that was present in the original tissue sample.

The basic experimental design used in several developmental toxicology studies employing these techniques involved the treatment of up to six dams for each experimental group. From these dams, at least three different embryos were selected to serve as the source of either the neural tube or craniofacial structures. The remaining embryos in the litter were available for histopathology or immunohistochemical analyses. This approach provided no less than 18 aRNA probes which were hybridized to slot blots in order to provide sufficient data to statistically analyze the changes in gene expression secondary to teratogen exposure *in utero*.

One such study involved an examination of gene expression in the branchial arches of mouse embryos that had received retinoic acid exposure during critical periods of morphogenesis. Retinoic acid (RA) in the form of isotretinoin (Accutane, Hoffman-LaRoche Laboratories; Nutley, NJ) is a potent human teratogen associated with a number of significant craniofacial congenital malformations.[60-61] For the most part, these malformations involve craniofacial derivative structures from the first and second branchial arches. In order to better appreciate how altered gene expression may contribute to the observed embryopathy, pregnant LM/Bc mice received a single, 5 mg/kg i.p. injection of all-*trans* RA on gestational day (GD) 8.5. Both the first and second branchial arches were removed from control and RA-treated embryos on GD 10, GD 10.5, and GD 12.0. For these studies, 40 candidate genes were examined for the effect of the drug treatment on their coordinate expression. Using the aforementioned IST/aRNA approach, it was clear that *in utero* exposure to RA resulted in significant differences in the expression level of the nicotinic acetylcholine receptor subunit a (*NaChR*), transforming growth factor β2 ($TGF\beta_2$), type 1 cellular retinoic acid binding protein (*CRBP-1*), retinoic acid receptor γ (*RARγ*) and cAMP response element binding protein (*CREB*). Specifically, there was a significant upregulation of *CREB* expression within the first branchial arch of the RA-treated embryos at GD 10 when compared to the control embryos or to the level of expression within the RA-treated second arch. Similarly, there was a significant upregulation of *RARγ* expression within the second branchial arch of the RA-treated embryos at GD 10 when compared to the control or to the level of expression within the RA-treated first arch.[59] *CRBP-1* expression also differed significantly between the first and second branchial arches, although the drug treatment did not significantly alter their endogenous patterns of expression.[59] At GD 10.0 and 10.5, the RA treatment significantly reduced *NAChR* expression in the first branchial arch when compared to the controls; however, as the fetuses developed, the situation reversed and first branchial arches from the RA-treated embryos had a significant upregulation of *NAChR* expression when compared to controls at GD 12.0.[59] Finally, RA treatment significant reduced the expression of $TGF\beta_2$ in the first branchial arch at GD 10.5. There was significantly more expression of this gene in the second as opposed to the first branchial arch at GD 12.0, although the RA-treatment significantly downregulated the gene at this timepoint.[59] The net effect of the observed changes may reflect a unique response within the second branchial arch that not only increases the amount of RA that is transported into the cells, but also serves to enhance the activity of $RAR\gamma$, a nuclear receptor for RA that directly controls transcriptional regulation. Together with the downregulation of $TGF\beta_2$, which is directly involved in cell cycle progression, cellular differentiation, adhesion, migration, and extracellular matrix production,[62] the changes in gene expression may have a functional role in compromising the normal developmental processes involved in craniofacial morphogenesis and result in the observed congenital malformations.

In a similar study examining changes in embryonic gene expression following chronic exposure to teratogenic concentrations of the anticonvulsant drug phenytoin (Dilantin, Parke, Davis; Ann Arbor, MI), inbred SWV mouse embryos were harvested on GD 9.5 and 10.0 and frozen in liquid nitrogen. Eight-micron thick mid-sagittal sections were cryostat cut at –30°C and mounted on diethylpyrocarbonate-treated, gelatin subbed slides. The substrate cDNA obtained by IST/aRNA from these sections was amplified into amplified antisense RNA and used as a probe to screen a panel of 20 candidate cDNA clones that were considered important regulators of craniofacial and neural development. This investigation revealed that chronic *in utero* phenytoin exposure produced a significant downregulation in the expression of $TGF\beta_1$, *NT3* (neurotrophin 3), *Wnt-1*, *NaChR*, and the voltage-sensitive calcium channel gene in GD 9.5 embryos. For all five of the downregulated genes, the pattern of expression at GD 9.5 most closely resembled control embryo gene expression at GD 10.0.[63] As a result of the teratogenic drug treatment, several important developmental genes were altered during a critical period of morphogenesis. For those changes which result in altered patterns of functional protein levels, there may be significant embryonic consequences.[67]

For both of the aforementioned investigations, it was clear that while changes in individual gene expression was no doubt important, the data supports the hypothesis that no one single gene is responsible for the teratogen-induced changes in normal development. It was readily apparent that it was the coordinate regulation of several molecules, each of which by itself may not be sufficient to elicit detectable embryological changes, but when combined together, they were capable of producing the adverse phenotypic changes associated with the prenatal phenytoin embryopathy.[64-65]

Acknowledgments

This work was supported in part by grants from the Public Health Service, ES 07165 and DE 11303 to Dr. Richard H. Finnell and DA 00199 to Dr. Scott A. Mackler. The authors gratefully acknowledge the expert clerical assistance provided by Ms. Melissa Powell.

References

1. McBride, W.G., Thalidomide and congenital abnormalities, *Lancet,* ii, 1358, 1961.
2. Lenz, W. and Knapp, K., Thalidomide embryopathy, *Arch. Environ. Health,* 5, 100, 1962.
3. Smithells, R.W., Defects and disabilities of thalidomide children, *Br. Med. J.,* 1, 269, 1973.
4. Fraser, F.C. and Fainstat, T.D., The production of congenital defects in the offspring of pregnant mice treated with cortisone. A progress report, *Pediatrics,* 8, 527, 1951.
5. Biddle, F.G., Teratogenesis of acetazolamide in the CBA/J and SWV strains of mice. I. Teratology, *Teratology,* 11, 31, 1975a.
6. Biddle, F.G., Teratogenesis of acetazolamide in the CBA/J and SWV strains of mice. II. Genetic control of the teratogenic response, *Teratology,* 11, 31, 1975b.
7. Nau, H., Rating, D., Koch, S., Häuser, I., and Helge, H., Valproic acid and its metabolites: placental transfer, neonatal pharmacokinetics, transfer via mother's milk and clinical status in neonates of epileptic mothers, *J. Pharmacol. Exp. Ther.,* 219, 768, 1981.
8. Finnell, R.H., Bennett, G.D., Karras, S.B., and Mohl, V.K., Common hierarchies of susceptibility to the induction of neural tube defects by valproic acid and its 4-propyl-4-pentenoic acid metabolite, *Teratology,* 38, 313, 1988.
9. van Gelder, R., von Zastrow, M.E., Yool, A., Dement, W.C., Barchas, J.D., and Eberwine, J.H., Amplified RNA synthesized from limited quantities of heterogeneous cDNA, *Proc. Natl. Acad. Sci. U.S.A.,* 87, 1663, 1990.
10. Eberwine, J.H., Yeh, H., Miyashiro, K., Cao, Y., Nair, S., Finnell, R.H., Zettel, M., and Coleman, P., Analysis of gene expression in single neurons, *Proc. Natl. Acad. Sci. U.S.A.,* 89, 3010, 1992.
11. Eberwine, J.H., Spencer, G., Miyashiro, K., Mackler, S.A., and Finnell, R.H., cDNA synthesis *in situ*: methods and applications, in *Methods in Enzymology,* Part G, *Recombinant DNA,* Wu, R., Ed., Academic Press, New York, 1992, 80.
12. Saiki, R.K., Primer-directed enzymatic amplification of DNA with a thermostable DNA polymerase, *Science,* 239, 487, 1985.
13. Wang, A.M., Doyle, M.V., and Mark, D.F., Quantitation of mRNA by the polymerase chain reaction, *Proc. Natl. Acad. Sci. U.S.A.,* 86, 9717, 1989.
14. Eisenstein, B.I., The polymerase chain reaction, *N. Engl. J. Med.,* 322, 178, 1990.
15. Erlich, H.A., Gelfand, D., and Sninsky, J.J., Recent advances in the polymerase chain reaction, *Science,* 252, 1643, 1991.
16. Haase, A., Rezel, E., and Staskus, K., Amplification and detection of lentiviral DNA inside cells. *Proc. Natl. Acad. Sci. U.S.A.,* 87, 4971, 1990.
17. Liang, P. and Pardee, A.B., Differential display of eukaryotic messenger RNA by means of the polymerase chain reaction, *Science,* 257, 967, 1992.
18. Templeton, N.S., The polymerase chain reaction. History, methods and applications, *Diag. Mo. Pathol.,* 1, 58, 1992.

19. Botteman, C.D. and Sommer, S., PCR amplification of specific alleles: rapid detection of known mutations and polymorphisms, *Mut. Res.,* 288, 93, 1993.
20. Maniatis, T., Goodbourn, S., and Fischer, J.A., Regulation of inducible and tissue specific gene-expression, *Science,* 236, 1237, 1987.
21. Alberts, B., Bray, D., Lewis, J., Raff, M., Roberts, K., and Watson, J.D., *Molecular Biology of the Cell,* 3rd ed., Garland Publishing, New York, 1983, 409.
22. Sambrook, J., Fritsch, E.F., and Maniatis, T., *Molecular Cloning: A Laboratory Manual,* 2nd ed., Cold Spring Harbor Laboratory Press, Cold Spring Harbor, NY, 1989, 10.38.
23. Sive, H.L. and St. John, T., A simple subtractive hybridization technique employing photoactivatable biotin and phenol, *Nucleic Acids Res.,* 16, 10937, 1988.
24. Duguid, J.R., Rohwer, R.G., and Seed, B., Isolation of cDNAs of scrapie-modulated RNAs by subtractive hybridization of a cDNA library, *Proc. Natl. Acad. Sci. U.S.A.,* 85, 5738, 1988.
25. Herfort, M.R. and Garber, A.T., Simple and efficient subtractive hybridization screening, *BioTechniques,* 11, 598, 1991.
26. Miyashiro, K., Dichter, M., and Eberwine, J., On the nature and differential distribution of mRNAs in hippocampal neurites: implications for neuronal functioning, *Proc. Natl. Acad. Sci. U.S.A.,* 91, 10800, 1994.
27. Sambrook, J., Fritsch, E.F., and Maniatis, T., *Molecular Cloning: A Laboratory Manual,* 2nd ed., Cold Spring Harbor Laboratory Press, Cold Spring Harbor, NY, 1989, 7.16.
28. RNAzol™, Biotecz Laboratories, Friendswood, TX.
29. Chomczynski, P. and Sacchi, N., Single-step method of RNA isolation by acid guanidinium thiocyanate-phenol-chloroform extraction, *Anal. Chem.,* 162, 156, 1987.
30. Liang, P., Averboukh, L., and Pardee, A., Distribution and cloning of eukaryotic mRNAs by means of differential display: refinements and optimization, *Nucleic Acids Res.,* 21, 3269, 1993.
31. Finnell, R.H., Moon, S.P., Abbott, L.C., Golden, J.A., and Chernoff, G.F., Strain differences in heat-induced neural tube defects in mice, *Teratology,* 33, 247, 1986.
32. Callard, D. Lescure, B., and Mazzolini, L., A method for the elimination of false positives generated by the mRNA differential display technique, *BioTechniques,* 16, 1096, 1994.
33. Ausubel, F.M., *Current Protocols in Molecular Biology,* Suppl. 26, John Wiley & Sons, New York, 1994, 15.8.1.
34. Sambrook, J., Fritsch, E.F., and Maniatis, T., *Molecular Cloning: A Laboratory Manual,* 2nd ed., Cold Spring Harbor Laboratory Press, Cold Spring Harbor, NY, 1989, 7.39.
35. Sanger, F., Nicklen, S., and Coulson, A.R., DNA sequencing with chain terminating inhibitors, *Proc. Natl. Acad. Sci. U.S.A.,* 74, 5463, 1977.
36. Eberwine, J.H., Valentino, K.L., and Barchas, J.D., *In Situ Hybridization in Neurobiology: Advances in Methodology,* Oxford University Press, Oxford, 1994.

37. Bauer, D., Muller, H., Reich, J., Riedel, H., Ahrenkiel, V., Warthoe, P., and Strauss, M., Identification of differentially expressed mRNA species by an improved display technique (DDRT-PCR), *Nucleic Acids Res.,* 21, 4272, 1993.
38. Ayala, M., Balint, R.F., Fernandez-de-Cossio, M.E., Canaan-Haden, L., Larrick, J.W., and Gavilondo, J.V., New primer strategy improves precision of differential display, *BioTechniques,* 18, 842, 1995.
39. Tokuyama, Y. and Takeda, J., Use of ^{33}P-labeled primer increases the sensitivity and specificity of mRNA differential display, *BioTechniques,* 18, 424, 1995.
40. Liang, P., Zhu, W., Zhang, X., Guo, Z., O'Connell, P., Averboukh, L., Wang, F., and Pardee, A.B., Differential display using one-base anchored oligo-dT primers, *Nucleic Acids Res.,* 22, 5763, 1994.
41. Donohue, P.J., Alberts, G.F., Hampton, B.S., and Winkles, J.A., A delayed-early gene activated by fibroblast growth factor-1 encodes a protein related to aldose reductase, *J. Biol. Chem.,* 269, 8604, 1994.
42. Trentmann, S., Van der Knaap, E., and Kende, H., Alternatives to ^{35}S as a label for the differential display of eukaryotic messenger RNA, *Science,* 267, 1186, 1995.
43. Reeves, S., Rubio, M., and Louis, D., General method for PCR amplification and direct sequencing of mRNA differential display products, *BioTechniques,* 18, 18, 1995.
44. Mou, L., Miller, H., Li, J., Wang, E., and Chalifour, L., Improvements to the differential display method for gene analysis, *Biochem. Biophys. Res. Commun.,* 199, 564, 1994.
45. Sager, R., Anisowicz, A., Neveu, M., Liang, P., and Sotiropoulou, G., Identification by differential display of alpha 6 integrin as a candidate tumor suppressor gene, *FASEB J.,* 7, 964, 1993.
46. Liang, P., Averboukh, L., Keyomarsi, K., Sager, R., and Pardee, A., Differential display and cloning of messenger RNAs from human breast cancer vs. mammary epithelial cells, *Cancer Res.,* 52, 6966, 1992.
47. Liang, P., Averboukh, L., Zhu, W., and Pardee, A., Ras activation of genes: *Mob-1* as a model, *Proc. Natl. Acad. Sci. U.S.A.,* 91, 12515, 1994.
48. Zhang, L. and Medina, D., Gene expression for specific genes associated with mouse mammary tumor development, *Mol. Carcinog.,* 8, 123, 1993.
49. Hsu, D., Donohue, P., Alberts, G., and Winkles, J., Fibroblast growth factor-1 induces phosphofructokinase, fatty acid synthase and Ca^{2+}-ATPase mRNA expression in NIH 3T3 cells, *Biochem. Biophys. Res. Commun.,* 1483, 1993.
50. Zimmermann, J. and Schultz, R., Analysis of gene expression in the preimplantation mouse embryo: use of mRNA differential display, *Proc. Natl. Acad. Sci. U.S.A.,* 91, 5456, 1994.
51. Aiello, L.P., Robinson, G.S., Lin, Y., Nishio, Y., and King, G., Identification of multiple genes in bovine retinal pericytes altered by exposure to elevated levels of glucose by using mRNA differential display, *Proc. Natl. Acad. Sci. U.S.A.,* 91, 6231, 1994.
52. Douglas, J., PCR differential display identifies a rat brain mRNA that is transcriptionally regulated by cocaine and amphetamine, *J. Neurosci.,* 15, 2471, 1995.

53. Nishio, Y., Aiello, L., and King, G., Glucose induced genes in bovine aortic smooth muscle cells identified by mRNA differential display, *FASEB J.*, 8, 103, 1994.
54. Thelu, J., Burnod, J., Bracchi, V., and Ambroise-Thomas, P., Identification of differentially transcribed RNA and DNA helicase-related genes of *Plasmodium falciparum*, *DNA Cell Biol.*, 13, 1109, 1994.
55. Diaculangan, D., Chawia, A., Boak, A., Kagan, H., and Lazar, M., Retinic acid prevents downregulation of *RAS* recision gene/lysyl oxidase early in adipocyte differentiation, *Differentiation,* 58, 47, 1994.
56. Burn, T., Petrovich, M., Hohaus, S., Rollins, B., and Tenen, D., Monocyte chemoattractant *protein-1* gene is expressed in activated neutrophils and retinoic acid-induced human myeloiid cell lines, *Blood,* 84, 2776, 1994.
57. Bennett, G.D., Mohl, V.K., and Finnell, R.H., Embryonic and maternal heat shock responses to known teratogenic insults, *Reprod. Toxicol.,* 4, 113, 1990.
58. Stemple, D.L. and Anderson, D.J., Isolation of a stem cell for neurons and glia from the mammalian neural crest, *Cell,* 71, 973, 1992.
59. Taylor, L.E., Bennett, G.D., and Finnell, R.H., Altered gene expression in murine branchial arches following *in utero* exposure to retinoic acid, *J. Craniofacial Genet. Dev. Biol.,* 15, 13, 1995.
60. Lammer, E.J., Chen, D.T., Hoar, R.M., Agnish, N.D., Benke, P.J., Braun, J.T., Curry, C.J., Fernhoff, P.M., Grix, A.W., Lott, I.T., Richard, J.M., and Sun, S.C., Retinoic acid embryopathy, *N. Engl. J. Med.,* 313, 837, 1985.
61. Lammer, E.J., Preliminary observations on isotretinoin-induced ear malformations and pattern of formation of the external ear, *J. Craniofacial Genet. Dev. Biol.,* 11, 292, 1991.
62. Wrana, J.L., Attisano, L., Carcamo, J., Zentella, A., Doody, J., Laiho, M., Wang, X.F., and Massague, J., TGFb signals through a heteromeric protein kinase receptor complex, *Cell,* 71, 1003, 1992.
63. Musselman, A.C., Bennett, G.D., Greer, K.A., Eberwine, J.H., and Finnell, R.H., Phenytoin-induced alterations of early embryonic gene expression, *Reprod. Toxicol.,* 8, 383, 1994.
64. Finnell, R.H., Abbott, L.C., and Taylor, S.M., The fetal hydantoin syndrome: answers from a mouse model, *Reprod. Toxicol.,* 3, 127, 1989.
65. Finnell, R.H., Yerby, M.S., and Nau, H., General principles: teratogenicity of antiepileptic drugs, in *Antiepileptic Drugs*, 4th ed., Levy, R.H., Mattson, R.H., and Meldrum, B.S., Eds., Raven Press, New York, 1995, 209.

Appendix: Protocols for Nucleic Acid Amplification Techniques

1. *In Situ* Transcription From Fixed Sections

1. Place glass slides of embryo sections on 3-mm paper in a petri dish (150 × 15 mm). Moisten paper with DEPC-treated water.
2. Make a rubber cement moat around each of the embryo sections.
3. Add the hybridization mixture (below) to the sections (enough to cover the tissue section, in these examples ~25 µl per section).

   Hybridization Buffer (make fresh):

   112.5 µl of 20× SSC

   262.5 µl of water

   25 µl of deionized formamide

   Oligo dt-T7 primer stock solution (20 to 100 ng/µl). Add 1µl per section.

   [AAACGACGGCCAGTGAATTGTAATACGACTCACTATAGGCGC(T_{24})]
4. Incubate at room temperature overnight in the closed humidification chamber.
5. Remove the hybridization mixture. Let sit in Copeland jar containing 2× SSC for 15 minutes at room temperature. Repeat wash with new 2× SSC for 15 minutes, then perform two 60-minute washes in 0.5× SSC.
6. Re-cement the moats and add 18 to 20 µl of the IST reaction mixture (below) to each section.

   IST Reaction Mixture (for 50 µl):

   5 µl of 10× IST buffer (500 m*M* Tris (pH8.3), 60 m*M* $MgCl_2$, 1.2 *M* KCl)

   3.75 µl of 0.1 *M* Dithiothreitol

   1.25 µl of dATP (10 m*M*)

   1.25 µl of dGTP (10 m*M*)

   1.25 µl of dTTP (10 m*M*)

   1 µl of ^{32}P-labeled α dCTP (800 µCi per 25 µl reaction); (optional): if ^{32}P α dCTP is not included, add 1.25 µl of dCTP.

   2 µl of reverse transcriptase (Seikagaku, to a final concentration of 1 U per 1 µl) water to final volume of 50 µl
7. Float dish in 37 to 41°C waterbath for 1 to 2 hours. Do not bump or mix fluids.
8. Wash in 2 l DEPC-treated 2× SSC for 6+ hours at room temperature.
9. Wash in 2 l DEPC-treated 0.5× SSC for 6+ hours at room temperature.
10. If radioactivity was added, air dry slides, remove moats, tape slides to cardboard backing, and expose to X-ray film to confirm that ^{32}P incorporation occurred. A short exposure (less than 30 minutes) is all that should be needed.
11. Pipette each embryo section off slide with 20 µl of fresh 0.5 *N* NaOH/0.1% SDS solution and place each section in a separate eppendorf tube. To each tube, add 20 µl of 1 *M* Tris (pH 8.0).

12. Perform a phenol:chloroform:isoamyl (25:24:1) extraction and ethanol precipitate with 1/10 volume of 3 *M* sodium acetate and 4 μg of glycogen.
13. Resuspend in 10 μl of water.
14. Add 2 μl of Klenow buffer (1× = 50 m*M* Tris, pH 7.0; 50 m*M* NaCl, 5 m*M* $MgCl_2$)

    1 μl of dATP (10 m*M*)

    1 μl of dGTP (10 m*M*)

    1 μl of dTTP (10 m*M*)

    2 μl of ^{32}P dCTP (radiation optional); If ^{32}P α dCTP is not included, add 1 μl of dCTP

    2 μl of water

    1 μl of T4 DNA polymerase (1 unit per μl)

    1 μl of Klenow (2 U/μl)
15. Incubate at 14°C for 4 hours to overnight.
16. Add 81 μl of 1× S1 nuclease buffer (30 m*M* NaOAc, pH 4.5; 50 m*M* NaCl, 1m*M* zinc acetate).
17. Add 1 μl of S1 nuclease enzyme (1 to 3 U/μl).
18. Incubate at 37°C for exactly 3 minutes.
19. Perform another phenol:chloroform:isoamyl (25:24:1) extraction followed by an ethanol precipitation using 4 μg glycogen as a carrier.
20. Resuspend the pellet in 4.5 μl of water. Add the following:

    1 μl of Klenow buffer

    1 μl of dATP (10 m*M*)

    1 μl of dGTP (10 m*M*)

    1 μl of dCTP (10 m*M*)

    1 μl of dTTP (10 m*M*)

    1 μl of T4 DNA polymerase (1 U/μl)
21. Incubate at 37°C for 20 minutes.
22. Drop dialyze for at least 3 hours using a Millipore 0.025 μm VS type filter against water, or use a spin column. Now ready for aRNA amplification protocol.

2. In Situ *Transcription From Fresh Embryos*

1. Remove desired embryonic tissue (e.g., neural tube or branchial arch) and place into 50 μl of hybridization buffer.

    Fresh Tissue Hybridization Buffer (can be stored at 4°C; make fresh due to the DTT):

    50 m*M* Tris, pH 8.3

    6 m*M* $MgCl_2$

    120 m*M* KCl

    1 mg/ml Digitonin

    5 m*M* Dithiothreitol

2. Place the tube on ice and sonicate the tissue with a microtip for approximately 20 seconds. This helps to permeabilize the tissues. Heat sample 80°C 5 minutes. Place on ice 5 minutes.
3. To 50 µl of sonicated sample containing tissue add the following:
   - 0.5 µl RNasin®
   - 1 µl of oligo dT-T7 primer (20 to 100 ng/µl)
   - 1 µl of dATP (10 m*M*)
   - 1 µl of dGTP (10 m*M*)
   - 1 µl of dTTP (10 m*M*)
   - 1 µl of dCTP (10 m*M*)
   - 1 µl of reverse transcriptase (25 U/µl)

Optional: *If you are having a problem with your substrates, try 2 µl of reverse transcriptase.*

4. Incubate at 41°C for 1 to 2 hours in a heat block.
5. Extract with phenol:chloroform:isoamyl (25:24:1) and ethanol precipitate with sodium acetate and 4 µg glycogen.
6. Heat at 95°C for 5 minutes
7. Resuspend pellet in 5 µl water.
8. To resuspended pellet, add the following:
   - 2 µl Klenow buffer
   - 1 µl dATP (10 m*M*)
   - 1 µl dGTP (10 m*M*)
   - 1 µl dTTP (10 m*M*)
   - 1 µl dCTP (10 m*M*)
   - 0.8 to 1 µl T4 DNA polymerase (1 U/µl)
   - 0.4 to .5 µl Klenow (2 U/µl)
   - 7.8 µl water
9. Incubate at 14°C for at least 4 hours or at room temperature for 3 hours.
10. Add 20 µl 10× S1 nuclease buffer and 160 µl water. Add 1 µl S1 nuclease enzyme (1 to 3 units per µl).
11. Incubate at 37°C for exactly 3 minutes.
12. Extract with phenol:chloroform:isoamyl (25:24:1) and ethanol precipitate with sodium acetate and 4 µg glycogen.
13. Resuspend in 4.5 µl of water. Add the following to blunt end:
   - 1 µl of Reaction 2 buffer
   - 1 µl of dATP (10 m*M*)
   - 1 µl of dGTP (10 m*M*)
   - 1 µl of dCTP (10 m*M*)
   - 1 µl of dTTP (10 m*M*)
   - 0.5 µl of T4 DNA polymerase (1 U/µl)

14. Incubate at 37°C for 20 minutes.
15. Add 190 ml water. Perform another phenol:chloroform:isoamyl alcohol extraction and ethanol precipitation.
16. Pass sample over a spin column (Millipore #UFC3 LTK 00) and resuspend in 20 μl water.
17. Now ready for aRNA amplification.

3. aRNA Amplification Procedure

1. For a 25-μl reaction add the following:

   5 μl 5× amplification buffer (200 m*M* Tris (pH 7.5), 30 m*M* $MgCl_2$, 50 m*M* NaCl, 10 m*M* spermidine)

   5 μl cDNA from IST protocol

   1 μl CTP (100 μ*M*)

   1 μl UTP (10 m*M*)

   1 μl GTP (10 m*M*)

   1 μl ATP (10 m*M*)

   1 μl 0.1 *M* Dithiothreitol

   7 μl water

   1 μl RNasin® (20 U/μl)

   2 μl ^{32}P CTP (3000 Ci per mmol)

   25 μl total
2. Label strip of 1-mm Whatman paper with sample numbers (before and after). Remove 0.5 μl of aRNA reaction mixture and spot in "before" position on Whatman paper.
3. Add 2 μl of T7 RNA polymerase (50 U/μl or higher if available) to amplification mixture.
4. Incubate at room temperature for 1 hour, then incubate at 37°C for 3 hours.
5. Remove 0.5 μl of reaction and place in "after" position on Whatman paper. Place 1-mm paper in 10% TCA for 5 to 10 minutes, decant TCA, and repeat for another 5 to 10 minutes. Let paper air dry, then count on scintillation counter.

Note: *aRNA can be used as a probe (see probe protocol) or continued for additional amplifications (step 6).*

6. Add 25 μl of water to total volume of 50 μl. Perform a phenol:chloroform:isoamyl (25:24:1) extraction followed by an ethanol precipitation with sodium acetate and 4 μg glycogen.
7. Resuspend pellet in 10 μl of water. Heat denature for 5 minutes at 80°C then quickly place on ice.
8. For cDNA synthesis, add the following to 10 μl of sample:

   1 μl random primer (hexamer) mix (10 to 20 ng/μl)

   2.5 μl of I.S.T. buffer (10×)

1 µl of 0.1 *M* Dithiothreitol
1 µl of dATP (10 m*M*)
1 µl of dGTP (10 m*M*)
1 µl of dCTP (10 m*M*)
1 µl of dTTP (10 m*M*)
5 µl of water
1 µl of reverse transcriptase (25 U/µl; final concentration 1 U/µl)
0.5 µl of RNasin®

9. Incubate at 37 to 41°C for 1 to 2 hours.
10. Add water to bring volume up to 50 µl. Add NaOH (final concentration at 0.2 *M*) and let stand at room temperature for 10 to 15 minutes. Extract and precipitate as before.
11. Resuspend dried pellet in 10 µl of water.
12. Heat denature at 95°C for 3 minutes. Place on ice for 5 minutes. Quick spin in centrifuge.
13. Add:

    1 µl of oligo-dT-T7 primer (50 to 100 ng/µl)
    1 µl of dATP (10 m*M*)
    1 µl of dGTP (10 m*M*)
    1 µl of dCTP (10 m*M*)
    1 µl of dTTP (10 m*M*)
    2.5 µl of Klenow buffer
    6.3 µl of water
    0.8-1 µl of T4 DNA polymerase (1 to 3 U/µl)
    0.4 to .5 µl of Klenow (2 U/µl)
    Final volume 25 µl

14. Incubate at 14°C for at least 2 hours (3+ hours are fine) or 3 hours at room temperature.
15. Bring volume to 50 µl with an additional 25 µl of water. Extract and precipitate sample as before.
16. Add the following to blunt end:

    1 µl of Reaction 2 buffer
    1 µl of dATP (10 m*M*)
    1 µl of dGTP (10 m*M*)
    1 µl of dCTP (10 m*M*)
    1 µl of dTTP (10 m*M*)
    0.5 µl of T4 DNA polymerase (1 U/µl)

17. Incubate at 37°C for 20 minutes.
18. Add 190 ml water. Extract as before.
19. Run sample over a spin column (Millipore #UFC3 LTK 00) to remove unincorporated nucleotides.

Hybridization

20. Label nylon membrane blots (Zeta Probe, BIO RAD) on non-DNA side with pencil. Place each blot in a separate 50-ml centrifuge tube.
21. Add 150 μl of prehybridization solution per cm^2 Zeta-Probe GT membrane (50% formamide, 0.12 *M* Na_2HPO_4, pH 7.2, 0.25 *M* NaCl, 7% (w/v) SDS) to each tube. Prehybridize blots for 5 minutes at 43°C.
22. Heat denatured riboprobe at >75°C for 5 minutes. Put on ice for 5 minutes.
23. Add probes to centrifuge tubes containing prehybidized blots. Incubate overnight at 43°C or optimized temperature.
24. Be sure to keep 2 μl of aRNA for running on a denaturing gel.

Washes

25. Dispose of hybridization solution and probe in radioactive waste.
26. Perform a quick wash in 2× SSC. Dispose of radioactive 2× SSC.
27. Wash blot for 1 hour in Wash #1 (2× SSC + 0.1% SDS) using at least 350 μl solution per cm^2 Zeta-Probe GT membrane for each wash. Temperature (37 to 65°C) may vary with desired stringency. Be sure to preheat your wash solutions to desired temperature before washing.
28. Wash for 1 hour in Wash #2 (0.5× SSC + 0.1% SDS).
29. Wash for 2 hours in Wash #3 (0.1× SSC + 0.1% SDS).
30. Wash in 2× SSC in Tupperware briefly as a final rinse.
31. Blot dry and wrap in plastic wrap. Expose blot to X-ray film or an imager.

4. Differential Display PCR

The following protocol is based upon the original method of Liang and Pardee.[17] Modifications and notes are indicated.

1. Start with 2 μg DNA-free total RNA.
2. Divide the RNA into four 250-ng portions for reverse transcription as indicated below. Save the remaining RNA at –80°C.

    250 ng DNA-free total RNA

    2 μl 10× transcription buffer[a]

    1 μl Dithiothreitol (100 m*M*)

    2 μl T_{12}MN (10 μ*M*)[b]

    1.6 μl dNTP mix (250 μ*M* each)

    0.5 μl RNasin®

    x μl sterile water

    18 μl total volume

3. Heat samples to 65°C for 5 minutes followed by 37°C for 10 minutes.
4. Add 2 μl AMV reverse transcriptase (Seikagaiku; Bethesda, MD) and continue incubation for 1 to 2 hours at 41°C.[c]

5. Heat inactivate the reverse transcriptase by incubation for 5 minutes at 95°C.
6. PCR amplify the cDNA as follows:[d]

   9.2 μl sterile water

   2 μl 10× PCR buffer

   1.6 μl dNTP (25 μ*M*)

   2 μl arbitrary primer (2 μ*M*)[e]

   2 μl T_{12}MN (10 μ*M*)[f]

   2 μl cDNA from R.T. reaction

   1 μl α ^{35}S-dATP (1200 Ci per mmol)[g]

   0.2 μl AmpliTaq (Perkin-Elmer)

   20 μl total volume
7. The PCR conditions are as follows:

   94°C 1 minute denaturing

   40°C 2 minutes annealing

   72°C 1 minute extension

   Repeat 40 cycles followed by 5-minute hold at 72°C.

   Cover samples with 25 μl mineral oil.
8. Mix 3.5 μl of each PCR sample with 2 μl of gel loading dye[h] and incubate for two minutes at 80°C prior to electrophoresis.
9. Run the PCR products on a 6% denaturing acrylimide gel such that samples amplified with identical primer pairs are loaded in adjacent lanes. Run the gel at 60 W until the xylene cyanol dye reaches the bottom of the gel.
10. Following electrophoresis, blot the gel on 3M paper, cover with plastic wrap and dry at 80°C on a vacuum gel drier.
11. Align the gel on X-ray film by hole punching or marking with radioactive ink such that the film can be realigned after exposure.
12. Develop film and identify differentially expressed cDNAs. CDNAs can be removed by simply cutting the acrylamide gel with a razor blade.
13. Place the isolated gel fragment in a microcentrifuge tube containing 100 μl water and allow to rehydrate for 10 minutes. Cover the tube with parafilm and boil for 15 minutes.
14. Centrifuge the tube for a few minutes to collect condensation, then transfer the supernatant to a new microcentrifuge tube.
15. To the supernatant add:

   10 μl 3*M* sodium acetate

   5 μl glycogen (10 mg/ml)

   450 μl 100% ethanol
16. Precipitate on dry ice for 30 minutes. Centrifuge for 10 minutes and wash pellet with 200 μl 85% ethanol.
17. Redissolve pellet in 10 μl water.

18. Reamplify the isolated cDNA using the same PCR primers used to generate it on the initial DD gel. The PCR conditions are the same as those described earlier, except for the dNTP concentration. For reamplification, 250 μ*M* is recommended instead of a 25 μ*M* dNTP mix.
19. Verify the isolated cDNA by Northern blot analysis or slot blotting. In addition, the cDNA may be further characterized by cloning, sequencing, and *in situ* hybridization.

Modifications and Notes

[a]10× transcription buffer: 500 m*M* Tris (pH 8.3), 1.2 *M* KCl, 100 m*M* $MgCl_2$.

[b]Each reverse transcription reaction should contain a different $T_{12}MN$ primer ($T_{12}MA$, $T_{12}MC$, $T_{12}MG$, $T_{12}MT$). Because the $T_{12}MT$ primer may *resolve as a smear,* it normally is not used. To avoid this problem, one-base anchored primers with restriction sites[40] may be used.

[c]MMLV reverse transcriptase also may be used.

[d]The conditions listed are compatible with the Gene Amp™ kit (Perkin-Elmer) and the RNAmap™ kit (GeneHunter), although Ampli Wax gems[26] and other methods are available.

[e]Although the original protocol involved the use of a select group of 10mers, additional improvements in primer selection have been introduced.[36] In addition, the arbitrary primers may be end-labeled for greater sensitivity[39] and elongated to target genes with specific binding domains.[41]

[f]The $T_{12}MN$ anchored primer used in PCR must be the same as that used in the respective reverse transcription reaction to generate the cDNA.

[g]Because the use of ^{35}S in PCR results in a volatile decomposition product,[42] ^{33}P has been recommended as the isotope of choice for differential display.

[h]Gel loading dye: 95% formamide, 10 m*M* EDTA (pH 8.0), 0.09% xylene cyanole FF, and 0.09% bromphenol blue.

Part II

Protein Identification

I. Introduction

A hallmark of the complex morphogenetic changes which accompany embryonic development is the alteration of both tissue form and pattern. Underlying these alterations are continually changing patterns of protein expression which provide a rich variety of differentiation markers by which normal development can be followed. These developmentally expressed proteins may also serve as biomarkers for the assessment of developmental toxicity. In addition, beyond their mere utility as marker proteins, many such molecules themselves also play a fundamental role in mechanically driving morphogenesis. Central to the identification and analysis of these important molecules is a variety of biochemical and immunological tools and techniques for the analysis of developmentally regulated protein expression. This chapter provides an overview of methods of protein analysis, including protein separation by chromatographic and electrophoretic techniques, and the production and application of antisera and monoclonal antibodies.

The methods described are, in general, universally applicable to analysis of protein expression in any developing system. However, the development and utilization of these methods are often trial-and-error empirical processes which must be adapted to each investigator's particular situation.

It is a truism that there are as many variations on the details of these methods as there are laboratories using them. Because space is limited, this review, therefore, is meant to serve as an introductory guide, and the cited references should be used as a rich source of greater theoretical and methodological detail for the interested reader who wishes to apply these techniques. A single source that provides a wealth of theoretical and practical information on all the methods discussed in this review is the excellent laboratory manual by Harlowe and Lane.[1]

Methods for development of antisera and monoclonal antibodies described here involve the experimental use of laboratory animals. Thus, investigators using these methods should consult their institutional animal care and use committees for procedural guidelines which will assure compliance with local and federal regulations. Care of animals such as rabbits and mice, on which many of the techniques discussed in this chapter rely, is discussed in the text by Harkness and Wagner.[2]

Finally, for all of the techniques described, there is no amount of reading that can substitute for hands-on experience. Consultation and collaboration with experienced colleagues is strongly recommended for investigators who wish to introduce these techniques into their own laboratories.

II. General Principles of Protein Structure

A. Primary Structure

Fundamental to the biochemical and immunological analysis of protein variations during normal and abnormal development is a general understanding of protein structure. This is also relevant to the development and application of immunological tools, such as antibodies, for protein analysis. Antibodies are themselves proteins, and the assays and techniques described in this chapter have at their core the protein-protein interactions which occur between antibodies and their antigens. While nonproteinaceous macromolecules may, of course, be antigenic and elicit an immune response, such applications are beyond the scope of this chapter. Proteins are composed of peptide-bonded covalently linked linear arrays of amino acids, and the linear sequence of amino acids in any given protein (polypeptide) is referred to as its primary structure. The incredible variety of proteins, from those found in bacteria to those of humans, is based upon variations in the sequence of the same 20 common amino acids. Normally included in a description of the primary sequence is also the location of any internal disulfide bonds formed between cysteine residues. Even small variations in amino acid sequence can result in sufficiently different protein properties to permit their distinction by biochemical and immunochemical methods.

B. Higher Order Structure

The structure of a given protein depends not only on the primary sequence of its amino acids but also on the internal noncovalent interactions of amino acids making up different parts of the protein as a result of charge or hydrophobic interactions. Interactions of closely neighboring residues leads to secondary structures, such as the formation of α-helices or β-pleated sheets, while the interactions of more distantly localized residues results in tertiary folding into more complex structures. Finally, two or more polypeptide chains may interact noncovalently to form homo- or heteromultimers, a level of organization which is referred to as quaternary structure. These various types of interactions result in changes in size, shape, charge, and hydrophobicity of proteins which again can be exploited in the analysis of their expression patterns.

C. Size and Shape

The size of proteins commonly encountered in the analysis of developing systems may range from small polypeptides such as cytokines, composed of perhaps a few dozen amino acids with a combined mass of a few thousand Daltons, to monumental proteins of several thousand amino acids and a mass of hundreds of thousands of Daltons. The size differences between proteins can be exploited by a number of biochemical methods such as gel electrophoresis or column chromatography to resolve complex mixes of proteins which then can be detected by chemical or immunological methods. Proteins of similar mass can have very different shapes due to differences in their folding patterns. Even under native conditions, proteins occur in a range of forms from long, straight rods to short, stubby globules. Furthermore, the shape of a protein can be changed, as when globular proteins extend into rods when complexed with detergents during electrophoresis. These shape changes again provide a means of separating protein mixtures as during molecular sieve chromatography.

D. Charge and Isoelectric Point

The overall electrical charge of a protein depends on the summed charges of the amino acids it is composed of. Neutral and positively and negatively charged amino acids all occur in proteins under native conditions. Furthermore, since the charge of an amino acid can depend on conditions such as pH, the overall charge of a protein will be dependent on the environment in which it exists. As in other properties of proteins, charge differences can be exploited to separate protein mixtures, as during ion-exchange chromatography. The isoelectric point of a protein is the pH at which it has an effective neutral charge, by virtue of a balance between the ionization states of its component amino acids. The procedure known as isoelectric focusing (IEF) exploits this fact to also resolve complex mixes of proteins. IEF is often used in combination with other methods, as in the first step of two-dimensional gel electrophoresis, with the second step based upon protein size differences.

E. Hydrophobicity

Individual amino acids differ in their polarity and, hence, their water solubility. Likewise, a protein generally will be hydrophilic or hydrophobic, depending on its amino acid composition. A large protein has many distinct domains that vary in this property, and the higher order structure of a protein is largely determined by folding into distinct hydrophilic and hydrophobic regions. Detergents are often required to solubilize membrane-associated hydrophobic portions of proteins and to keep them in solution. Such proteins must be solubilized before they can be subjected to the separation techniques discussed in this chapter. As with other variations in protein properties, hydrophobicity can be exploited as a separation method, as during hydrophobic column chromatography.

F. Posttranslational Modifications

The various amino acids along the protein backbone provide a rich variety of chemically modifiable sites which can further contribute to protein variation. Such modifications may be as chemically simple as the addition of a phosphate group or as complicated as the addition of a huge carbohydrate moiety. Such groups alter the charge and bulk of proteins and, hence, their structure and function. Indeed, even a small modification can have profound effects on these aspects of protein behavior. In addition, such differences can again be used as a basis for separation and detection of differentially modified proteins, as with lectin affinity chromatography for glycoproteins or site-specific antibodies for phosphorylated proteins.

G. Affinity and Activity

No protein functions in a vacuum, and the basis for the structural and catalytic activity of proteins within a biological system is their greater or lesser binding affinity for a variety of ligands ranging in size from small ions to enormous macromolecules. Indeed, such affinities of antibodies for their target antigens form the basis for the application of the immunochemical detection methods described below. These methods of solid-phase absorption can be exploited to take advantage of such affinities for the further separation and analysis of proteins from complex mixes. Finally, the function of a protein itself can be exploited in the separation and detection of protein differences, through the development of activity-based assays. For example, enzymatic catalysis may be followed biochemically and provides a very sensitive method. However, antibody-based inhibition of cellular function can also be used in the development of protocols for the identification of unknown proteins the existence of which can only be inferred from their activity.

III. Principles of Protein Separation and Analysis

The analysis of protein structure and expression requires the application of a number of analytical methods which take advantage of the above properties of proteins to permit their separation and detection. These methods vary in their resolution and sensitivity and include both chromatographic techniques (gel permeation, ion exchange, hydrophobicity, and affinity) and electrophoretic techniques (one- and two-dimensional gels). The general properties and virtues of each is discussed below, and good general discussions can be found in the references by Deutscher[3] and Robyt and White.[4]

A. Resolution and Sensitivity

When considering the application of protein separation techniques, two major criteria for selection are the resolution and sensitivity of the methods. Resolution is important to consider, as proteins that differ greatly in their properties (say, size or charge) may be separated by less stringent methods. Sensitivity becomes an issue, especially if relatively small quantities of proteins are to be analyzed, as is often the case in studies of embryonic systems. Polyacrylamide gel electrophoresis in the presence of sodium dodecyl sulfate detergent (SDS-PAGE) may resolve even large proteins that differ by as little as 1 kD in molecular weight, while gel permeation chromatography has a much smaller resolving power. However, the latter technique has the virtue of handling large amounts of protein and can be performed with proteins in their native conformation. SDS-PAGE results in the denaturation of proteins. Isoelectric focusing may separate proteins differing by as little as 0.01 pH units, while ion-exchange chromatography again has a much lower resolving power. However, the latter method again has the virtues of working with large amounts of protein and being relatively simple to carry out. Once separated by any of these methods, the ability to detect the resolved proteins will also depend on the detection method. A general chemical stain for proteins, such as Coomassie blue, can detect individual protein bands on a gel which each contain 0.1 to 1.0 μg protein. The more sensitive silver stain method is an order of magnitude more sensitive, while radioactive tracing methods or immunodetection methods can give yet another order of magnitude sensitivity, down to about 0.01 to 0.001 μg protein.

B. Column Chromatography

Gel permeation chromatography, also known as size exclusion or gel filtration chromatography, separates proteins according to their size and shape. The essence of the technique is the passage of a protein mixture over a long, narrow glass column containing small beads that have both a controlled diameter as well as a controlled pore size within the beads. As the protein mixture passes through the bed of beads, the largest proteins in the mixture will be entirely excluded from the beads and will pass most quickly through the column. Proteins of intermediate size will spend some time inside the pores of the beads, and some time outside the beads. The smallest of proteins will spend most of their time winding their way through the many small pores of the beads. In this fashion, the time it takes a protein to pass through the column will be proportional to the size of the protein. An advantage of this technique is that it usually can be carried out under mild conditions preserving the structure and activity of the protein. However, resolution is relatively poor, and this method works best for proteins which differ considerably in size. Careful consideration of the proteins to be separated, along with choice of column design as well as bead packing material, can lead to optimization of the method. Factors to be considered in the choice of beads for packing the column include the exclusion limit, fractionation range, particle shape, and size. Beads can be made of polyacrylamide, agarose, or dextran, and the choice of material will depend on whether one is separating small peptides or large macromolecular complexes. Literature provided by the manufacturer must be consulted to aid in these decisions. In general, a twofold difference in molecular weight is often required for an efficient separation using this method. For example, this method is excellent for separating IgM from other types of immunoglobulins, due to the large size of the former. A good resource for gel filtration methodologies is the text by Fischer.[5]

Ion-exchange chromatography is a relatively simple technique that is especially useful for separating large quantities of proteins. Combined with salting-out precipitation using ammonium sulfate, this is the method of choice for purification of polyclonal antibodies from serum or of monoclonal antibodies from hybridoma conditioned medium or ascites fluid. The principle of the technique is the use of a column packed with a solid phase composed of small beads containing charged groups whose bound ions may be exchanged with charged proteins. Commonly used matrices consist of cellulose, dextran, or agarose modified to contain DEAE (anion-exchanging diethtylaminoethyl) or CM (cation-exchanging carboxymethyl) sidegroups. Modified cellulose is most commonly used, and, as acidic proteins are more common than basic proteins, DEAE cellulose is used more often than CM cellulose. The binding of proteins to the matrix depends on the charge of the protein and is, therefore, a function of the buffer used to suspend the protein and equilibrate the column. Thus, the buffer's ionic strength and pH determine whether a given protein binds to or passes through the column. Unlike gel filtration, which requires long thin columns, ion-exchange can be done in relatively short columns or even as a batch procedure. While the general procedure involves the four steps of column equilibration, sample loading, sample elution, and column regeneration, the detailed procedure and buffer conditions must be tailored to the individual protein of interest. Ion-exchange chromatography is discussed by Robyt and White.[4]

Affinity chromatography, a very potent method of protein separation, has great potential for selectivity that relies on the specificity of protein-ligand interactions. The solid-phase ligands may range from specific small organic or inorganic molecules to large proteins themselves, such as antibodies. In this latter case, the method is referred to as immunoaffinity chromatography and is discussed further below. Passage of a mixture of proteins over an immunoaffinity column results in the binding of the specific protein to its antibody ligand, while other proteins pass through. The target protein may then be eluted by changing the pH or ionic strength of the buffer solution, since protein-antibody binding is highly dependent on the charge and shape of the proteins. In theory, this technique can be used to isolate a protein in pure form in a single step. In practice, however, this method is often used in combination with the other chromatographic techniques described above. An excellent reference for affinity methods is the book by Dean.[6]

C. Gel Electrophoresis and Electroblotting

Gel electrophoresis is probably the single most powerful method for the analysis of complex protein mixtures when one considers both sensitivity and resolution. Electrophoresis refers to the phenomenon by which a charged molecule, such as a protein, will migrate when subjected to an electric field. The rate of migration will depend on the field strength, the charge on the molecule, and the mass of the molecule, as well as the effects of any environmental resistance to migration. All of these factors come into play in establishing conditions for protein gel electrophoresis and can be manipulated to enhance the resolution and separation of proteins in complex mixtures. Many variations on the technique exist, and detailed background on both theory and application can be found in the text by Hames and Rickwood.[7]

The SDS-PAGE (sodium dodecyl sulfate polyacrylamide gel electrophoresis) system is the most commonly encountered in the research laboratory. The use of acrylamide polymer as the solid anticonvective support has the advantages of optical clarity, chemical stability and inertness, and easily controllable pore size on the order of protein dimensions, providing a molecular sieving effect. Thin gel slabs cast between glass sheets are used as opposed to glass rods. Controlling pore size by the extent of acrylamide cross-linking permits the resolution of a wide range of proteins, and the gradient method described here extends this capability beyond that obtained with a single concentration of acrylamide. The dissociating discontinuous buffer system provides an excellent level of protein solubility and resolution of proteins into tightly focused bands. While not described here, an additional level of resolution can be obtained through the use of two-dimensional gel electrophoresis, in which a protein mixture is first resolved according to isoelectric point prior to separation by SDS-PAGE. Two dimensional electrophoretic methods are discussed in detail by Dunbar[8] and by Hames and Rickwood.[7]

It is possible to probe proteins within an intact SDS-PAGE slab gel with various ligands following electrophoresis, but diffusion and pore size limit the convenient use of this approach to ligands much smaller than antibodies. This problem has been solved through the advent of protein electroblotting, in which an applied electric field moves proteins from the gel horizontally onto an apposed sheet of nitrocellulose paper, which has a high affinity for proteins. The transferred proteins on the nitrocellulose sheet is referred to as a Western blot, to distinguish it from Northern blots of mRNA and Southern blots of DNA (there really is a Dr. Southern). The Western blot provides a replica of the gel banding pattern on the surface of the nitrocellulose sheet, in which form the proteins are easily accessible for probing with antibodies in the procedure known as immunoblotting. A method for Western immunoblotting is described below and is discussed further by Harlowe and Lane.[1]

IV. Nature of the Immune Response

Organisms have developed defensive mechanisms for dealing with invasion by foreign organisms or molecules. As part of the response to these antigens, the immune system produces antibodies, proteins which recognize and bind to antigens at sites called epitopes. The antibodies produced as a defense against antigens are also extremely useful analytical reagents, since one can take advantage of their specificity and affinity for the isolation and identification of proteins. The following discussion defines several of these key concepts of antigen, epitope, and antibody structure; antibody-antigen interactions; clonal selection of B cells (the lymphocytes that make antibodies); and the differences between monoclonal and polyclonal antibodies. Excellent general references on immunology and related methods are those by Harlowe and Lane,[1] Weir,[9] and Roitt.[10]

A. Antigens

The specific molecules that elicit an immune response and which are bound by antibodies are called antigens. The antigencity of a given protein refers to its relative ability to cause an immune response when introduced into a host animal. When a protein is used to inoculate an animal in order to raise antibodies via the host's immune response, the protein is referred to as an immunogen. The immunogenicity (or antigencity) of a given protein will depend on the physical properties of the protein, as well as the genetic background and prior environmental experience of the host animal. As a practical matter, proteins vary widely in their antigencity in a given host, and different hosts, even within a species, will be found to vary widely in their response to a given antigen. Therefore, the generation of antibodies for experimental use becomes, to some extent, an empirical exercise. Thus, the use of multiple animals in an experimental series is recommended due to host variability.

B. Antibodies

Antibodies are specialized proteins synthesized by B lymphocytes. Antibodies are produced via genetic recombination from multiple genes, resulting in a variety of antibody subclasses (IgG, IgM, IgA, IgE, and IgD) which differ somewhat in their structure and function, as well as in their synthesis at different stages of an immune response. The structure of a typical antibody molecule (an IgG) is illustrated in Figure 6.1. Antibodies are Y-shaped heterotetramers composed of two light chains of about 25,000 Da mol wt and two heavy chains of about 50,000 Da mol wt. These four chains are held together by disulfide bridges to give a single-linked tetramer of about 150,000 Da in size. Reduction of the protein by an agent such as mercaptoethanol leads to separation of the chains, which can be done in a controlled fashion. In addition, selective proteolysis of the IgG with enzymes such as papain or pepsin results in the fragmentation of the antibody into distinct Fab (antigen-binding) and Fc (crystallizable) regions that have antigen recognition and complement-fixing activities, respectively. The ease of production of such fragments differs between Ig subclasses and species types, but such fragments often are useful since they are monovalent and no longer cross-link target antigens.

C. Epitopes

Antibodies usually recognize small portions of a protein, since these antigens can be quite large compared to the antigen-binding region of an antibody. The portion of the antigen which the antibody recognizes and binds to is called an epitope. With respect to proteins discussed here, an epitope may involve up

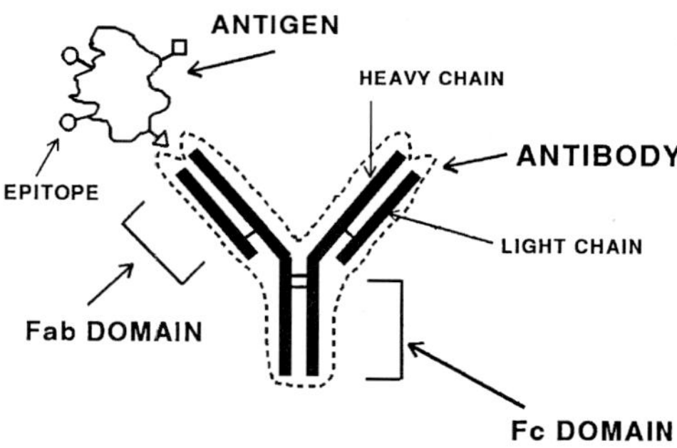

FIGURE 6.1
The diagram illustrates the structure of immunoglobulin subtype G (IgG), the variety found at highest concentration in serum. The IgG molecule is composed of two heavy and two light chains linked together by disulfide bridges. The Fab (antigen binding fragment) domain includes the antigen binding sites, contributed to by regions of both the heavy and light chain proteins, which are encoded by highly variable regions of their corresponding genes. This variation gives rise to the large repertoire of possible antibodies. The Fc (crystallizable fragment) domain is more constant in structure, hence the ability to crystallize it when purified. Also illustrated is a model protein antigen, with several distinct epitopes available for antibody recognition and binding, one of which is recognized by this particular antibody.

to one to two dozen amino acid residues, forming regions of the antigen which bind the antibody through multiple noncovalent bonds. The different bonds work cooperatively, and even small changes in portions of this epitope can result in large differences in binding. This contributes to the great specificity of antibody-antigen reactions.

D. Antibody Specificity

While the above sensitivity of antibody-antigen binding to changes in antigen structure gives rise to the phenomenon of antibody specificity, cross-reactions of antibodies may occur to multiple antigens which differ greatly in structure but share one or more epitopes. This is more of a problem with monoclonal antibodies, which recognize best a single unique epitope, since the multiple epitopes typically involved in a monospecific polyclonal antiserum become averaged out over the target antigen.

E. Clonal Selection

Antigen recognition and antibody synthesis by lymphocytes requires a complex chain of events involving different cell types. While a discussion of the cellular and molecular complexities of the immune response are far beyond the scope of this chapter, Figure 6.2 provides a brief summary of the key

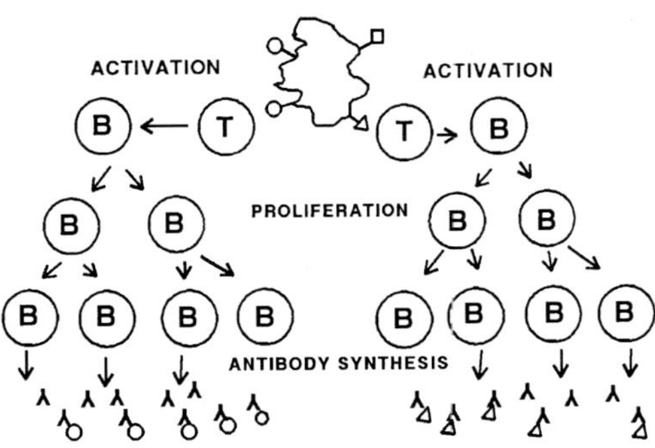

FIGURE 6.2
Clonal selection of B cells. Both B and T cells interact as part of the immune system network, and they cooperate in the maturation of B lymphocytes into antibody secreting cells. Any one activated B cell and its clonally derived mature progeny secrete only one type of antibody protein molecule. A complex antigen such as that illustrated here possesses a number of potential epitopes which can each stimulate the activation of a different clone of B cells. Thus, immune serum, even if resulting from immunization with a highly purified antigen, usually contains a mixture of many antibodies and is referred to as a polyclonal antiserum. Hybridoma production depends on the ability of B cells to be hybridized with myeloma cells (B-cell tumors), resulting in the production of stable cell lines derived from a single activated B cell and producing a single type of monoclonal antibody.

elements of B-cell activation, proliferation, and antibody synthesis and secretion. While T lymphocytes serve to regulate immune responses or directly kill foreign cells, it is the B lymphocytes which produce antibodies. Somatic recombinant genetic mechanisms occurring among antibody genes in lymphocyte precursors allow these cells to respond potentially to millions of different antigens. However, any one mature B lymphocyte produces and secretes only one type of antibody. Thus, stimulation of the immune system with a protein composed of multiple epitopes results in the activation and selection of multiple clones of B lymphocytes, each making a different antibody type.

F. Polyclonal Antibodies

With the above in mind, it is possible to understand the different origins and experimental utility of polyclonal vs. monoclonal antibodies. Polyclonal antibodies are mixtures of multiple antibody types found in the serum of immunized animals, hence the term antiserum. Depending on how the animal was immunized, the antibodies in an antiserum may recognize multiple antigens or multiple epitopes of a single antigen. Even when an animal is immunized with a highly purified protein antigen, because proteins are so large, potentially many epitopes may result in clonal selection of many B cells, resulting in a mixture of antibodies in the serum. Thus, when making antisera, antigen purity is critical, as garbage in will yield garbage out and a possibly useless antiserum without the desired specificity. Methods for polyclonal antiserum production are discussed below.

G. Monoclonal Antibodies

Monoclonal antibodies are produced *in vitro* by specialized cell lines called hybridomas, which result from the fusion of activated B lymphocytes taken from the spleen of an immunized animal and a B-cell tumor cell called a myeloma. The resulting hybridomas have desirable properties derived from both parent cells, the immortality of the myeloma (previously selected so that it does not secrete its own antibodies), and the specific antibody-secreting ability of the B cell. Each monoclonal antibody recognizes a single epitope, since it is synthesized by cells clonally derived from a single activated B cell. Since highly purified clonally-derived hybridoma cell lines can be selected and propagated in culture, hybridomas provide a theoretically endless source of highly specific antibodies, even if the original immunogen itself was not highly purified. Methods for monoclonal antibody production are discussed below.

V. Techniques for Production of Polyclonal Antibodies

The overall scheme for polyclonal antibody production is diagrammed in Figure 6.3. The following discussion summarizes the major steps and considerations in designing a protocol for polyclonal antiserum production. An excellent single source for additional information is the extensive laboratory manual by Harlowe and Lane.[1]

A. Select a Host

All aspects of animal care and use must be approved by your host institution to assure compliance with local and federal law. Consult your local animal care and use committee prior to initiating experiments. For the vast majority of experimental laboratory situations, female New Zealand white rabbits are the animal of choice, since they are large enough to provide significant volumes of serum, but small enough to use small amounts of antigen for immunization and to be housed conveniently. Other animals such as mice, guinea pigs, hamsters, or chickens may be used in special situations. Farm animals can be used for commercial scale production. The following procedures assume the use of rabbits. It is suggested that three animals be used per protocol, as not all individuals will respond in the same manner. Prior to immunization, preimmune serum samples should be drawn and tested to be certain the animal does not produce antibodies giving spurious background reactions.

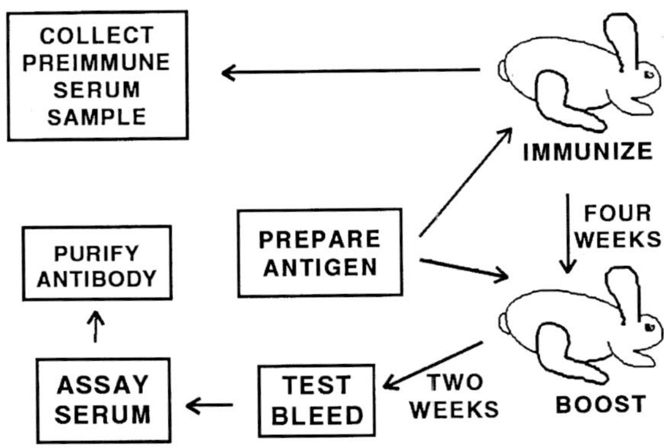

FIGURE 6.3
Production of polyclonal antibodies. Rabbits are the most common hosts for antiserum production. Following collection of pre-immune serum as a control, the animal is immunized to produce a primary immune response, mainly IgM. After a month, the animal is again immunized with antigen, and the resulting secondary response leads to production of higher concentrations of serum antibodies, mainly IgG. Boosting and antiserum collection can recur at monthly intervals, often resulting in production of high titer antiserum. Since the antiserum is polyclonal in nature, the use of a highly purified antigen is critical to obtaining a specific antiserum.

B. Prepare the Antigen

In general large proteins make good antigens since they potentially contain a wide variety of epitopes. Small peptides less than a few thousand Daltons in size may be problematic. The main concern for antigen preparation in polyclonal antibody production is purity of the protein immunogen. A purification scheme must be worked out for each specific situation, using any appropriate combination of protein chromatographic purification methods discussed earlier. As a final step, you may consider preparative scale gel electrophoresis, directly excising bands from the gels for immunization. Rabbits can tolerate macerated polymerized acrylamide gel fragments rather well.

C. Immunize the Host Animal

Remember to take pre-immune serum samples to save for control assays. One milligram of protein per immunization would be ideal, but even as little as 1 to 10 μg may be sufficient, since immunogenicity is highly variable from protein to protein. The most common route of immunization for proteinaceous immunogens is subcutaneous injection at 5 to 10 sites on the rabbit's back. This may be done directly if the protein is embedded in acrylamide gel fragments, but soluble proteins should be injected as an emulsion in an adjuvant such as Freund's for best results. Booster injections should be given at monthly intervals. Immunizing too often can induce tolerance, but animals remain primed for over a year. Following each booster immunization, high levels of circulating antibodies can persist for up to a month.

D. Collect Antiserum

It is recommended to take test bleeds of about 10 ml from the marginal ear vein of each rabbit 2 weeks after each booster immunization. Once a good titer of antibodies has been established, continue monthly boosts, collecting 25 ml of serum 2 weeks after each boost. Blood should be allowed to coagulate at 4°C overnight, and the serum should be separated from the coagulum.

E. Screen Serum for Antibodies

It is essential to plan your screening assays well in advance of beginning immunizations, so that pre-immune serum can be tested and you are ready when the rabbits are. Saving some purified antigen for this purpose helps, but any suitable assay for which antibody use is intended (e.g., Western blotting, ELISA, or immunohistochemistry) can be used. Several such assays are described later.

F. Purify and Store Antibodies

Antibodies are very stable proteins, and antiserum can be stored for some time at 4°C (with 0.02% NaN_3 as a preservative) or stored indefinitely if frozen at –20°C. Aliquot your samples for storage, as repeated freezing and thawing cycles can denature the antibodies. For many purposes, antiserum need not be purified further before use. If necessary, the simplest method for purifying antibodies is precipitation with 50% ammonium sulfate followed by ion-exchange chromatography on DEAE-cellulose.

VI. Techniques for Production of Monoclonal Antibodies

The overall scheme for monoclonal antibody production is diagrammed in Figure 6.4. This procedure assumes the availability of a fully equipped cell culture laboratory including a cell culture incubator, laminar flow hood, liquid nitrogen cell storage facility, cell culture microscopes, and other ancillary equipment and cell culture supplies. Without good cell culture technique, successful hybridoma production is unlikely. An excellent cell culture manual is that by Freshney.[11] The following discussion summarizes the major steps

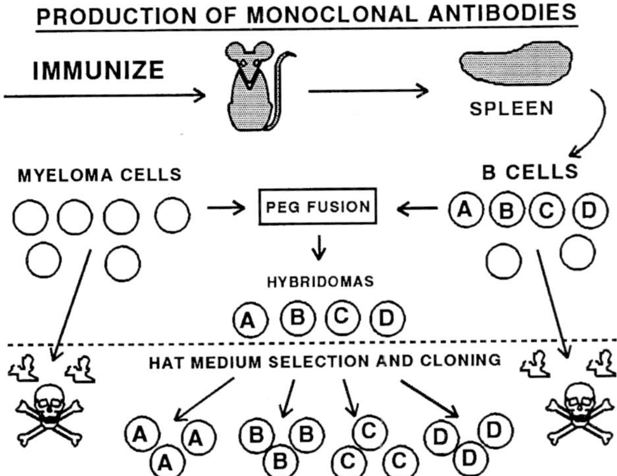

FIGURE 6.4
Mice are the most common hosts for monoclonal antibody production. Following an immunization schedule, the mice respond with production of a polyclonal antiserum. The animal is sacrificed, its spleen is removed, and the B cells so obtained are fused with myeloma (B cell tumor) cells using the membrane fusogen polyethylene glycol (PEG). Each resulting hybridoma possesses the specific antibody-secreting property of its parental B cell, as well as the immortality in culture of its parental myeloma. Unfused B cells normally do not persist long in culture and die. Unfused myeloma cells are killed by inclusion of hypoxanthine, aminopterin and thymidine (HAT) in the growth medium, which kills these cells as they lack the enzyme HGPRT which permits nucleotide biosynthesis even in the presence of these drugs. The hybridomas, which have inherited this enzyme from the parent B cell, do survive and may now be clonally selected in culture and screened for production of the desired specific monoclonal antibody.

and considerations in designing a protocol for monoclonal antibody production. Suggested references which contain a wealth of additional detail on theory and method are those by Goding[12] and Harlowe and Lane.[1]

A. Select a Host

Monoclonal antibodies are routinely produced only in mice and rats. Mice are the most common hosts, and females of the BALB/c strain are suggested. The available myeloma cell lines have been derived from BALB/c mice, and thus these mice can be used later for ascites production by intraperitoneal injection of hybridomas. Use at least three mice per immunization, and identify them individually for later screening and selection. Collect a small volume of pre-immune serum from each animal for later use as a control in initial screening assays.

B. Prepare the Antigen

Unlike polyclonal antibody production, antigen purity is not as big an issue, since unwanted antibodies can be screened out during the hybridoma selection.

However, it helps to use the purest protein antigen available that can be obtained using any of the chromatographic and electrophoretic procedures for protein purification described earlier.

C. Immunize the Mice

Purified protein antigens should be injected intraperitoneally as emulsion with complete Freund's adjuvant. A booster injection should be given two weeks later with antigen in incomplete Freund's adjuvant. Immunize with 1 to 50 μg protein per injection as available, in about 0.5 ml total volume. A final booster immunization should be given 3 days prior to cell fusion.

D. Test the Serum

Two weeks after each boost, serum samples obtained from tail bleeds should be tested against the antigen of interest by suitable assays such as ELISAs or dot blots (see below), since these assays require only small volumes of serum. It generally is not worth making hybridomas from mice if there is no detectable serum response against your antigen.

E. Perform the Hybridoma Fusion

A few days prior to fusion, split your myeloma cell cultures so they are in a logarithmic growth phase for best results during fusion. While a number of cell lines are available, the author has had success using the murine nonproducer, nonsecretor myeloma cell line called P3X63Ag8.653. Sacrifice the mouse by cervical dislocation and remove the spleen using sterile technique. Mince the spleen to release single cells, which should total about 10^8 per mouse spleen. Prepare a mixed suspension of the spleen cells and 10^7 myeloma cells in medium without serum, gently pellet the cells, remove excess medium, and (without disturbing the pellet) gently add 1 ml of 30% polyethylene glycol (the membrane fusing agent) in medium without serum to the pellet. Gently tap the tube to suspend the pellet, and incubate for 5 minutes at room temperature. Following incubation, wash the cells by pelleting and final resuspension in 100 ml complete selection medium (see below) with serum, and disperse the cells in 100-μl aliquots into a series of 10 96-well microtiter plates. Place in tissue culture incubator overnight.

F. Select the Hybridomas

Hybridoma selection relies in part on the fact that unfused B cells will survive only a few days in culture. In addition, the myeloma cell lines used for hybridoma production lack the enzyme HGPRT (hypoxanthine guanine phosphoribosyl transferase), and thus cannot utilize the enzyme-dependent salvage nucleotide synthesis pathway when the main pathway is blocked by the drug aminopterin. Inclusion of hypoxanthine (100 μ*M*), aminopterin (0.4 μ*M*), and thymidine (16 μ*M*) in the medium (HAT medium) thus results in the survival of only hybridomas which have the complementing immortality potential of the myelomas and the HGPRT enzyme of the spleen cells. Feed cultures with additional HAT medium with serum every 2 days.

G. Screen for Monoclonal Antibody Production

Within a week small colonies of surviving hybridomas should become visible in the microtiter plate wells. Their growth following fusion is quite rapid, and large numbers of colonies can be generated. Within a few days postfusion, tiny colonies will be visible, which by 1 week can be quite large and must be screened before cultures get out of hand. By 1 to 2 weeks postfusion, many colonies should be visible. The success rate of hybridoma production can vary widely between fusions, and a stringent selection method will allow you to reduce the number of cell cultures you must process as the colonies are expanded. It therefore cannot be emphasized too greatly that a simple, rapid, efficient, and proven screening procedure suitable for small volumes of conditioned medium (e.g., dot blot, ELISA, or immunohistochemistry) should be in place before hybridoma production. Use your assay to screen culture supernatants as colonies attain macroscopically visible dimensions.

H. Subclone and Store the Hybridomas

Depending on the percentage of surviving cells and the density of plating, there is the possibility that initial positive wells will contain more than one hybridoma clone. Positive clones should thus be subcloned at least twice by limiting dilution until every well is positive. Low density cultures survive better if done with a feeder layer of normal splenocytes prepared from an unimmunized mouse in the same way as for fusion. Once cloned, the subclass of the monoclonal antibody should be determined using specific antibodies (commercial kits are available). Long-term storage must be in liquid nitrogen using standard tissue culture cell line freezing methods.

I. Scale Up Antibody Production

Positive cell lines should be gradually expanded into larger culture vessels to scale up production. Cells should be weaned off of HAT medium gradually, by first culturing for several weeks in HT medium. Well conditioned tissue culture supernatants typically will contain 10 to 50 μg/ml of specific antibody. Intraperitoneal growth as an ascites tumor can yield 1 to 10 mg/ml of antibody.

J. Purify and Store the Monoclonal Antibodies

Conditioned medium or ascites fluid, or solutions of purified antibodies, can be stored for long periods frozen at –20°C. Monoclonal antibodies can be purified by using either a combination of ammonium sulfate precipitation and ion exchange chromatography or commercially available affinity chromatography supports.

VII. Techniques for Gel Electrophoresis and Western Blotting

A. Gel Electrophoresis

The theory and application of protein gel electrophoresis were discussed above and have been discussed extensively by Hames and Rickwood.[7] The following is a standard protocol for running one-dimensional SDS-PAGE gels suitable for detection of proteins and as the first step in Western immunoblotting, as diagrammed in Figure 6.5.

Electrophoresis Materials

- Commercial gel unit and power supply
- Acrylamide solution: 30 g acrylamide and 0.8 g bisacrylamide in 100 ml water
- Resolving gel buffer: 1.5 *M* Tris, pH 8.8
- Stacking gel buffer: 0.5 *M* Tris, pH 6.8
- Running gel buffer: 3.03 g Tris, 14.4 g glycine per l, 0.1% SDS
- 10% SDS
- 0.5% TEMED (N,N,N′,N′-tetramethylethylenediamine)
- 50% glycerol
- 15% ammonium persulfate
- 1.5% ammonium persulfate
- Water-saturated sec-butanol
- Commercially prepared prestained molecular weight markers
- Fixation and staining solution: 0.25% Coomassie blue R250 dye in 50% methanol, 10% acetic acid, 40% water
- Destaining solution: 30% methanol, 10% acetic acid, 60% water

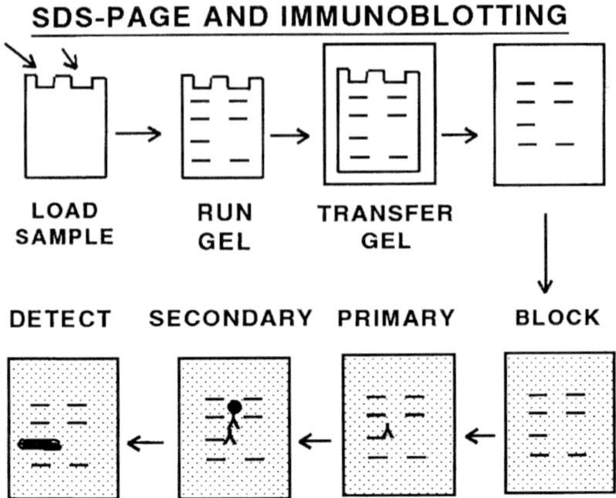

FIGURE 6.5
Electrophoresis and Western immunoblotting. Polyacryalmide gel electrophoresis in the presence of the detergent sodium dodecyl sulfate (SDS-PAGE) provides a potent method for resolving complex mixtures of proteins by migration in an electric field into single bands due the sieving effect of the slab gel. These protein patterns can be detected *in toto* by chemical stains such as Coomassie blue, and a difference between two samples can be detected if the proteins in question are present in sufficient quantity (as in the second panel). However, immunoblotting provides a highly sensitive alternative to detecting changes in protein patterns. The slab gel with the resolved proteins (having skipped the chemical staining step) are overlaid onto a sheet of nitrocellulose, which has a high capacity for binding proteins. A horizontally applied electric field removes the proteins from the slab gel and they become bound to the nitrocellulose sheet in the process known as Western blotting. After blocking excess binding sites with extraneous protein, a specific primary antibody can bind to the protein of interest. This primary antibody, in turn, can be detected with a secondary antibody that is covalently linked to an enzyme. In the final step, an enzyme reaction giving a colored, insoluble product leads to visualization of the band of interest.

Electrophoresis Procedure

- Take two glass plates, 6.25 × 7 in., and clean with 70% ethanol; dry and assemble in casting stands with 1.5-mm spacers
- In three vials, mix 5, 15, and 3.75% (stacking) acrylamide solutions, adding the APS last just before pouring the gel (see table below).

Component	5%	15%	3.75%
bis-Acryl 30:0.8	2.5 ml	7.5 ml	1.25 ml
Resolving gel buffer	3.9 ml	3.9 ml	—
Stacking gel buffer	—	—	2.5 ml
10% SDS	0.15 ml	0.15 ml	0.1 ml
H_2O	7.25 ml	1.05 ml	5.65 ml
50% glycerol	—	1.2 ml	—
0.5% TEMED	0.6 ml	0.6 ml	—
Straight TEMED	—	—	7.5 µl
15% APS	0.6 ml	0.6 ml	—
1.5% APS	—	—	0.5 ml

- Place the 5% and 15% solutions into chambers of gradient maker, add APS, and pour gel.
- Put a small amount of water-saturated butanol onto poured gel to make flat top and allow to gel.
- Pour off butanol, add APS solution to stacking gel solution, and add 5 ml onto top of resolving gel. Insert comb to form lanes, top off with stacking solution, and allow to harden.
- Remove the comb, assemble upper chamber, and fill with fresh running buffer.
- Load samples and prestained standards on top of gel.
- Place entire sandwich into gel box filled with running buffer (about 4 l) and run gel at 10 mA overnight or 30 mA for 5 to 6 hours.

The Coomassie blue stain described here is sensitive enough to detect 0.1 to 1 μg protein per band on the gel. Alternative techniques for total protein detection, such as silver staining, are at least an order of magnitude more sensitive. Note that once a gel is stained it cannot be used for a Western blot. Either skip this step or run duplicate gels, one of which will be stained for total protein and one which will be further processed for Western blotting.

- Remove gel from between glass plates.
- Place gel in dish and incubate overnight at room temperature in 250 ml staining solution.
- Remove stain and incubate gel with several changes of 250 ml destain solution until background is clear and protein bands are visible.

While stained gels can be stored for some time in destain solution, the dye eventually will be drawn off the proteins; therefore, gels should be documented by either photography or computerized image analysis. A number of apparatuses are commercially available for drying the original gels down on filter paper or cellophane sheets for permanent storage.

B. Western Blotting

The principles of Western blotting and immunostaining have been described above. This section describes the procedure for processing gels for western transfers and immunostaining following SDS-PAGE as described above and as diagrammed in Figure 6.5. Several protocols are also discussed by Harlowe and Lane[1] and by Dunbar.[8]

Western Protein Transfer and Immunoblotting Materials

- Commercially purchased transfer tank, cassette, sponges, and power supply
- 3-mm Whatman filter paper, 2 pieces 6.5 × 7 inches

- 0.45-μ pore size nitrocellulose paper, 5.5 × 6 inches
- Transfer buffer: 1× Tris-glycine, 20% methanol
- TBS with 2 m*M* calcium
- 200 ml 5% nonfat dry milk (10 g dry milk, 200 ml TBS/Ca)
- Primary antibody: any well characterized and specific antiserum or monoclonal antibody
- Secondary antibody, commercially prepared alkaline phosphatase conjugated goat antibodies specific for the species of primary antibody
- 50 ml AP detection buffer: 100 m*M* Tris, 100 m*M* NaCl, 5 m*M* $MgCl_2$, to which is added 12.5 ml NBT (nitro blue tetrazolium) solution (75 mg NBT in 300 μl H_2O, 700 ml dimethylformamide) and 75 μl BCIP (bromochloroindolyl phosphate) solution (50 mg BCIP in 1 ml dimethylformamide)

Western Protein Transfer Procedure

- After gel is finished, disassemble gel box and remove glass plates.
- Separate plates and assemble transfer cassette.
- Insert cassette into the transfer box so that piece of nitrocellulose is closest to the red positive terminal.
- Fill box with transfer buffer and run at 100 mA overnight or 200 mA for 4 to 5 hours.

Immunoblotting Procedure

- Remove the nitrocellulose, place in 100 ml 5% nonfat dry milk, and rock for 1 hour.
- Add appropriate predetermined amount of primary antibody directly to nonfat dry milk and rock overnight.
- Pour off primary antibody and wash membrane 3 times for 5 minutes with TBS/Ca.
- Place 100 ml fresh nonfat dry milk 5% on blot and add 20 μl of alkaline phosphatase conjugated secondary antibody (1:5000); rock 5 hours.
- Pour off secondary antibody mix and wash 3 times with TBS/Ca.
- Place 50 ml AP detection buffer onto blot and develop in dark for 30 to 60 minutes.

VIII. Techniques for Enzyme-Linked Immunosorbant ASSAY (ELISA)

A. Semiquantitative Dot Blot Immunoassays

The great power of Western immunoblots, as described above, is the combination of the resolving power of SDS-PAGE with the specificity and sensitivity of antibody-antigen reactions. The dot blot is a quick-and-dirty Western immunoblot in which the SDS-PAGE step is bypassed. Any protein-containing solution, from a crude tissue homogenate to a pure sample, can be spotted directly onto a piece of nitrocellulose paper, with all subsequent steps of blocking, washing, and immunodetection identical to that of a Western blot as earlier described. For the dot blot to be effective and interpretable, either the antigen or antibody (or both) must be clean and specific. For example, a monoclonal antibody of known specificity can be used to detect its target antigen in a spot of crude tissue homogenate. This provides a rapid, simple, and sensitive assay for a protein. While useful, dot blots are at best semiquantitative and the presence of cross-reactive material or background binding problems can confound interpretation of results. However, dot blots are often useful in situations where SDS-PAGE results in the denaturation of a protein such that it is no longer recognized by the available antibody. Thus, dot blots often can be used with nondenatured native proteins. Detection in a crude homogenate can be followed up with nondenaturing chromatographic methods, and resulting increasingly purified fractions can in turn be assayed by dot blotting.

B. Quantitative ELISA Immunoassays

Detection of proteins on dot blots and Western immunoblots involves the use of chromogen substrates which change from a colorless, soluble form into a colored, insoluble form, thus precipitating at the site of reaction, and resulting in development of a spot or band at the antigen location. ELISA assays provide a quantitative and sensitive version of the dot blot in which the antigen, immobilized in test tubes or wells of multi-well dishes, is detected in a method similar to the dot blot. However, the chromogen differs insofar as it changes from a colorless, soluble form to a colored, soluble form and remains in solution, as diagrammed in Figure 6.6. The optical density of the resulting colored solution then can be read in a spectrophotometer to quantify the reaction. Specially designed spectrophotometers are available which rapidly read large numbers of multi-well plates, thus permitting the automation and large-scale development of ELISA methods. ELISA techniques are discussed in detail by Harlowe and Lane.[1]

ELISA Reagents

- TBS: Tris-buffered saline, 10 m*M* Tris, 150 m*M* NaCl, pH 7.4
- Coating buffer: TBS
- Wash buffer: TBS containing 0.2% Tween 20 detergent
- Blocking solution: TBS containing 5% BSA
- Substrate solution: 1 mg/ml PNPP (paranitrophenylphosphate) in TBS
- Antigen: 1 μg/ml purified antigen or 1 mg/ml crude homogenate in TBS
- Primary antibody: any well characterized and specific antiserum or monoclonal antibody

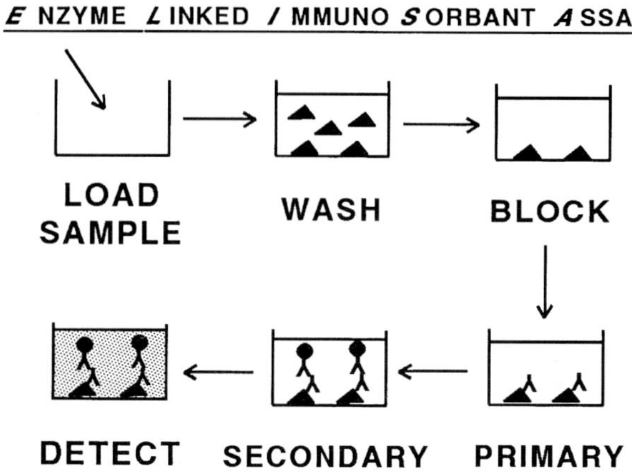

FIGURE 6.6
Enzyme-linked immunosorbant assay (ELISA) assays provide a sensitive and quantitative detection method for antigens. The sample containing the antigen of interest (triangles) is loaded into a series of wells in a polystyrene microtiter multi-well plate. After binding to the plastic, excess binding sites are blocked with the addition of extraneous protein. This is followed by incubation with a primary test antibody, which after washing is in turn reacted with a secondary detecting antibody conjugated to an enzyme. After again washing out unbound antibodies, a colorless substrate is added to the well, which turns into a colored but still soluble product. The wells can be conveniently read on a specially equipped spectrophotometer to quantify the reactions in each well.

- Secondary antibody: commercially prepared alkaline phosphatase-conjugated goat antibodies specific for the species of antibody used as the primary (e.g., rabbit for antisera, mouse for monoclonals)
- PVC plastic 96 multi-well microtiter plates

ELISA Procedure

- All steps can be done at room temperature.
- Prepare solution of antigen in TBS.
- Incubate 50 µl antigen solution per well for 1 hour.
- Remove antigen solution and block remaining binding sites by incubating with 200 µl blocking solution for 1 hour.
- Wash wells 3 times with 200 µl washing solution.
- Add 50 µl of primary antibody solution per well and incubate 1 hour (leave primary antibody out of some wells as a background control).
- Wash wells 3 times with 200 µl washing solution.
- Add 50 µl of secondary antibody solution per well and incubate 1 hour (leave secondary antibody out of some wells as a background control).
- Wash wells 3 times with 200 µl washing solution.
- Add 100 µl substrate solution to all wells.
- Incubate plate until yellow reaction color appears and read on spectrophotometer at 405 nm.

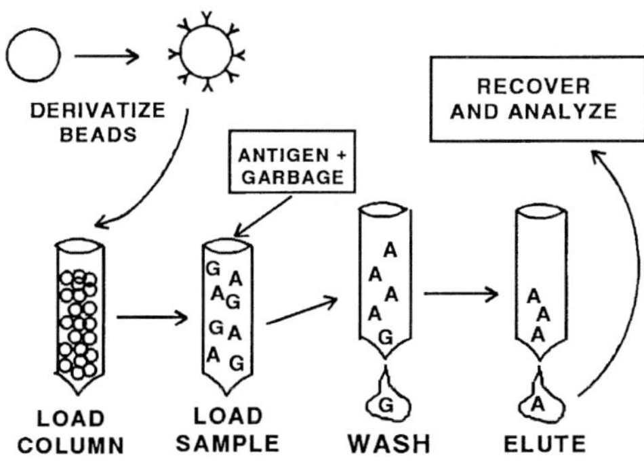

FIGURE 6.7
Immunoaffinity chromatography is a powerful separation technique which in theory permits the one-step isolation of highly purified proteins from even a crude mixture such as a tissue homogenate. The solid-phase absorbent can be custom-made by the covalent coupling of specific antibodies directly to small beads. Alternatively, the beads can be derivatized through binding of antibodies to commercially prepared beads already bearing linker antibodies. Following derivitization with specific antibody, the beads are loaded in a column. Passage of a complex mixture of solubilized proteins over the column results in binding of the antigen of interest (A) to the beads, while unwanted proteins (G for garbage) pass through. Following a wash to remove any remaining unbound protein, the antigen of interest can be recovered by eluting the column with a buffer of high ionic strength or extreme pH to disrupt the antibody-antigen interaction. The column can then be regenerated with loading buffer and can be reused multiple times.

IX. Immunoaffinity Techniques

Immunoaffinity chromatography and immunoprecipitation are very closely related solid-phase absorption techniques, both of which are used to purify antigens based upon their interaction with immobilized antibodies. The major difference in principle between these two methods is that in immunoaffinity chromatography, the beads containing the antibodies are immobilized in a column, while in immunoprecipitation the beads containing the antibody are sequentially pelleted and resuspended in a batch technique. The general schemes for these procedures are diagrammed in Figures 6.7 and 6.8, respectively, and include related steps of selecting an antibody support, derivatizing the beads with antibody, preparing the protein sample, binding the protein of interest to the immobilized antibody, removing unbound material, and recovering the protein of interest following dissociation from the antibody.

The details of either procedure will depend on the nature of the antibodies and antigens being utilized. In particular, the nature of the solid support can vary from covalently conjugated beads custom-made from commercially prepared activated beads to which you couple your own antibody, to noncovalently constructed sandwiches using a secondary antibody against your primary

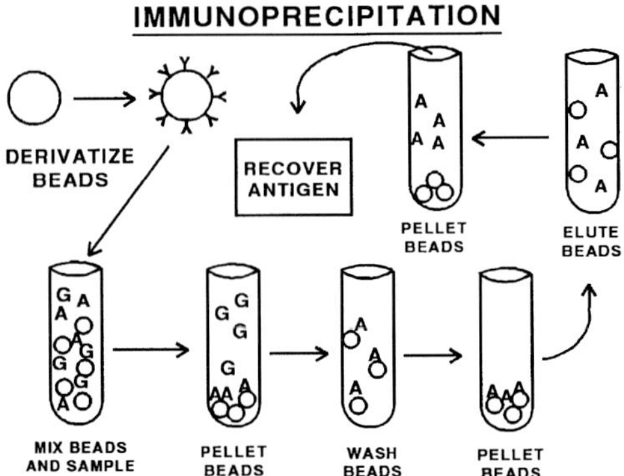

FIGURE 6.8
Immunoprecipitation is very similar to immunoaffinity chromatography except that it is done as a batch process. Beads derivatized with specific antibody are placed in a test tube with the solubilized protein mixture containing the antigen of interest. The antigen (A) binds to the beads, and unbound unwanted proteins (G for garbage) are removed by centrifugation and washing the beads. The antigen can be eluted from the beads by changing the buffer to one that gently disrupts the antibody-antigen binding, such as high or low pH (to recover the antigen) or SDS-PAGE sample buffer (for electrophoretic analysis).

antibody. In addition, a number of naturally occurring bacterial proteins (Protein A and Protein G) have a naturally high affinity for immunoglobulins. These proteins are often used to synthesize immunoaffinity beads useful in these procedures which can be commercially obtained. As a general guideline, the protocols below demonstrate particular variations of these methods suited to the specific examples. Additional detail and numerous variations on these methods which will further help the reader adapt protocols to their own situation can be found in the texts by Harlowe and Lane[1] and by Dean.[6]

A. Affinity Chromatography

Affinity Column Reagents

- Commercially prepared activated beads for covalent antibody immobilization via free amino groups
- 5 mg/ml purified antibodies in 0.1 *M* phosphate buffer, pH 7 (avoid buffers with amino groups, as these will react with the beads)
- Quenching buffer with amino groups: 1 *M* Tris, pH 7.5
- Columns for packing beads (2-ml bed volume)
- Well solubilized protein antigen, free of particulate matter, in equilibration buffer (containing detergent for membrane proteins; detergents must be used in all subsequent column steps to keep antigen in solution)
- Equilibration and wash buffer: 0.01 *M* phosphate, 0.15 *M* NaCl, pH 7.0
- Elution buffer: 100 m*M* thriethylamine, pH 11.5, or 100 m*M* glycine, pH 2.5

Immunoaffinity Column Procedure

Derivitization of Beads

Follow the manufacturer's suggestions for this step. A general procedure would involve incubating the beads with the antibody solution, mixing gently for several hours, washing the beads with phosphate buffer, then quenching any remaining activated sites with quenching buffer. The beads should then be washed with equilibration buffer, packed into the column, and stored at 4°C in equilibration buffer containing 0.02% NaN_3 as a preservative.

Immunoaffinity Purification

Load the antigen solution on to the column by repeated passing over the column. Several passes may be becessary for optimal binding. Wash the column with 10 column volumes of washing buffer. Elute the bound antigen with 10 column volumes high or low pH elution buffer, which will disrupt the antibody-antigen interaction. Collect 1 ml elution samples during this step into tubes containing 100 µl of appropriate neutralizing buffer. Recondition column with 10 column volumes of equilibration buffer with NaN_3 and store at 4°C. The column should be stable and reusable for months. The eluted fractions collected are ready for analysis by an appropriate assay for detection of the desired antigen.

B. Immunoprecipitation

Immunoprecipitation Reagents

- Primary antibody (in this example, NCD-2 rat monoclonal antibody against N-cadherin is the primary antibody)
- Secondary antibody (in this example, rabbit-anti-rat IgG, which acts as a capture antibody for the primary antibody)
- Solid support (in this example, recombinant rProtein G agarose beads, which have an excellent affinity for rabbit IgG but not for rat IgG)
- Triton immunoprecipitation buffer (Triton-IP buffer): TBS (0.01 *M* Tris, 0.15 *M* NaCl, pH 7.5), containing 5 m*M* EGTA, 1 m*M* $MgCl_2$, 1 m*M* PMSF, 2.5% Triton X-100
- SDS immunoprecipitation buffer (SDS-IP buffer): same as above, except 0.1% SDS instead of Triton X-100.
- Gel sample buffer: 0.0625 *M* Tris pH 6.8, 2% SDS, 10% glycerol, 5% 2-mercaptoethanol

Immunoprecipitation Procedure

- For each sample, mix 5 µl of rabbit-anti-rat IgG with 245 µl of Triton-IP buffer and mix well.
- For each sample, wash 50 µl of rProtein G agarose beads 2 times with 100 ml Triton-IP buffer.

- Add rabbit IgG solution to beads and incubate 1 hour with mixing at room temperature.
- Remove supernatant and wash IgG-conjugated beads 2 times with 100 μl Triton-IP buffer.
- Add 150 μl antibody solution (NCD-2 conditioned medium in this example) to bead suspension and incubate beads 1 hour with mixing.
- For control beads, use 150 μl Triton-IP buffer without antibody.
- Pellet beads, remove unbound antibody, and wash beads 2 times in 100 μl Triton-IP buffer.
- Add protein samples to beads (in this example, Triton-IP embryonic brain tissue extract containing solubilized N-cadherin) and incubate 4 hours with gentle mixing.
- Pellet beads, remove supernatant, wash beads 3 times with 300 μl Triton-IP buffer.
- For more stringent wash, use SDS-IP buffer. (For example, SDS-IP step will remove associated cytoskeletal proteins from bound N-cadherin, while use of Triton-IP buffer alone retains these proteins in the immunoprecipitate.)
- Pellet beads, remove supernatant, add 80 μl of 1× SDS-PAGE sample buffer onto beads.
- Mix beads; boil 5 minutes; mix, pellet, and load supernatant on SDS-PAGE gel.

References

1. Harlowe, E. and Lane, D., *Antibodies: A Laboratory Manual,* Cold Spring Harbor Laboratory Press, Cold Spring Harbor, NY, 1987.
2. Harkness, J.E. and Wagner, J.E., *The Biology and Medicine of Rabbits and Rodents,* 2nd ed., Lea & Febiger, Philadelphia, 1983.
3. Deutscher, M.P., *Methods in Enzymology,* Vol. 182, *Guide to Protein Purification,* Academic Press, San Diego, 1990.
4. Robyt, J.F. and White, B.J., *Biochemical Techniques: Theory and Practice,* Waveland, Prospect Heights, IL, 1987.
5. Fischer, L., *Gel Filtration Chromatography,* 2nd ed., Elsevier, New York, 1980.
6. Dean, P.D.G. et al., *Affinity Chromatography: A Practical Approach,* IRL Press, Oxford, 1986.
7. Hames, B.D. and Rickwood, D., *Gel Electrophoresis of Proteins: A Practical Approach,* IRL Press, Oxford, 1990.
8. Dunbar, B.S., *Two-Dimensional Electrophoresis and Immunological Techniques,* Plenum Press, New York, 1987.
9. Weir, D.M., *Handbook of Experimental Immunology,* Vol. 1, *Immunochemistry,* Blackwell Scientific, Oxford, 1986.
10. Roitt, I., *Essential Immunology,* Blackwell, Oxford, 1991.
11. Freshney, R.I., *Culture of Animal Cells: A Manual of Basic Technique,* Alan R. Liss, New York, 1987.
12. Goding, J.W., *Monoclonal Antibodies: Principles and Practice,* 2nd ed., Academic Press, San Diego, 1986.

Chapter 7

Immunohistochemistry

Philip E. Mirkes and Sally A. Little

Contents

0-8493-3342-3/97/$0.00+$.50

I. Introduction

One major area of interest in developmental toxicology is the effect of developmental toxicants on embryonic/fetal gene expression. Although there are several methods for assessing gene expression, e.g., Northern blot analysis, ribonuclease protection assay, reverse transcription-polymerase chain reaction (RT-PCR) and Western blot analysis, these methods are not readily adaptable to the analysis of gene expression in specific cells or tissues of the developing embryo or fetus. For this, two other methods are available, i.e., *in situ* hybridization (ISH) for the localization of gene transcripts (mRNA) and immunohistochemistry (IHC) for the localization of the corresponding protein. ISH is covered in Chapter 3; the focus of this chapter is IHC.

Although immunohistochemistry can be used to localize different antigens, including proteins, nucleic acids, carbohydrates, and a variety of other natural and synthetic compounds, this chapter will focus on the use of antibodies directed against protein epitopes to localize specific protein antigens in cells and tissues of mammalian embryos (mouse and rat). In addition, this chapter will be limited to methods applicable to immunohistochemistry at the light microscopic level. Nonetheless, it should be appreciated that similar approaches can be used to localize antigens to specific compartments within cells by employing methods for immuno-electron microscopy (see references in bibliography for detailed methods).

The primary purpose of this chapter is to provide a generic introduction to the immunohistochemical localization of specific proteins within cells/tissues of postimplantation mammalian embryos. A generic approach to this subject is taken because each antigen-antibody interaction is unique, and methods specific for the localization of one antigen may or may not be applicable to another antigen. In short, the localization of your antigen of interest will require careful modification of the protocols presented in this chapter in order to optimize a variety of parameters necessary for the successful localization of a specific protein. The ensuing discussion will assume that the investigator has access to relevant antibodies as well as other reagents and equipment listed in this chapter. Currently, antibodies specific to a wide variety of proteins can be obtained from three sources. First, antibodies and associated reagents often can be purchased from commercial companies (a list of some of the major suppliers is given in Appendix A). Second, one can obtain antibodies from individual researchers who have prepared them. Third, if these two sources are unavailable, one can prepare one's own antibodies. Preparation of antibodies is beyond the scope of this chapter; however, methods for the production of monoclonal or polyclonal antibodies can be found in references listed in the bibliography.

II. Methods and Techniques

A. Introduction

There are two basic approaches to the localization of proteins within cells/tissues of a postimplantation embryo, i.e., tissue-section immunohistochemistry, in which antibodies are applied to histological sections, and whole-mount immunohistochemistry, in which antibodies are incubated with permeabilized, but intact embryos. Tissue-section immunohistochemistry is the most common method in use to localize proteins within tissues. The advantages of this approach are (1) it is amenable to a wide variety of antibodies, and (2) it can be used to resolve localization patterns between adjacent cells/tissues. The primary disadvantage is that, unless laborious reconstruction algorithms are employed, the three-dimensional aspects of protein expression are difficult to obtain. Thus, it is relatively easy to determine that a particular tissue is expressing a particular protein; however, it is much more difficult to determine whether expression varies in different domains within a given tissue. Whole-mount immunohistochemistry, on the other hand, has as its main advantage the ability to capture the three-dimensional aspect of protein expression patterns. The main drawbacks of this approach are (1) the technique is laborious and time consuming, (2) not all antibodies that work in tissue sections work in whole mounts, and (3) the ability to resolve expression patterns with respect to intra- and intertissue levels is limited, although resolution can be increased by combining whole-mount immunohistochemistry with confocal microscopy or standard paraffin sectioning. With this as a general background, we will proceed to describe how we have used these two methods to localize proteins of interest within tissues of the mammalian embryo.

B. Tissue-Section Immunohistochemistry

In this section, we outline methods and provide protocols for the localization of proteins within tissues of the postimplantation mammalian embryo. This section begins with a discussion of methods of tissue preparation, proceeds to a discussion of important parameters associated with antibody recognition and binding to the antigen of interest, and concludes with a discussion of appropriate controls, the advantages and disadvantages of monoclonal vs. polyclonal antibodies, and the methods available to quantitate immunohistochemistry.

- Filter through Whatman #1.
- Add 20 g of paraformaldehyde.
- Heat to 60°C in fume hood.
- Add few drops of 1 *N* NaOH to dissolve.
- Cool and add 500 ml 2× PBS.

2. Fix embryos for 2 hours to overnight.
3. Follow standard embedding procedures.
4. Advantages:
   - Good fixative penetration
   - Excellent tissue morphology
5. Disadvantages:
   - Picric acid is toxic and requires special disposal procedures.
   - Picric acid solutions containing less than 10% water can be explosive.

2. Frozen Tissue

Although we have successfully visualized a variety of proteins in fixed, paraffin-embedded embryos, it is well known that some proteins cannot be visualized in this manner. The reasons for this failure are usually unknown but may include masking of the antigen by the fixative or loss of the antigen during the washing and dehydration in ethanols, hydrophobic interactions with xylene and paraffin, or heating during infiltration. Whatever the reason, at this point one turns to the use of frozen tissue and cryostat sectioning.

Below we outline general protocols for preparing tissue for cryosectioning. Note that either prefixation or postfixation is generally used with frozen tissues to prevent antigen dislocation and to preserve morphology. The key here is to use fixation conditions that achieve these objectives but do not completely mask the antigen of interest. One can achieve all three objectives by following the general protocols outlined below, but more importantly by obtaining assistance from an "expert" in cryostat sectioning.

No Prefixation

1. Place embryo in gelatin capsule.
2. Add OCT (mixture of polyvinyl alcohol, polyethylene glycol, and dimethyl benzylammonium chloride, available from Miles, Inc.; Elkhart, IN).
3. Freeze capsule in liquid nitrogen.
4. Store at –70°C for up to 1 year.

Prefixation

1. Fix embryos in formalin or paraformaldehyde for variable lengths of time.
2. Follow steps 1 to 4 above.

Postfixation

1. Follow steps 1 to 4 above.
2. Cut cryostat sections.
3. Fix tissue sections (acetone, formalin, methanol:acetic acid [3:1]).

D. Antibody Binding

At this point, the investigator will have appropriately fixed tissue sections in which one can localize the antigen of interest. In this section, we outline the basic factors to be considered in using antibodies to localize and visualize a specific antigen.

1. Detection Methods

The immunohistochemical localization of a specific protein begins with the application of the primary antibody to a tissue section. Once antigen antibody binding has occurred, it is necessary to visualize this complex. The next two sections outline the general procedures for visualizing antigen-antibody complexes.

Direct Method

Direct detection of antigen simply means that the primary antibody, directed against the antigen of interest, is coupled to a detection reagent (see below) and used to visualize the antigen (see Figure 7.1). The general protocol for direct detection of antigens is

1. Rinse slide containing tissue section with buffer.
2. Remove excess liquid from around section.
3. Incubate with blocking reagents (see Section II.D.5) and rinse.
4. Apply appropriately diluted primary antibody (see Section II.D.3). Incubate for minutes to hours.
5. Remove antibody; repeat steps 1 and 2.
6. Activate detection reagent (see Section II.D.2)
7. Rinse, counterstain, and coverslip.

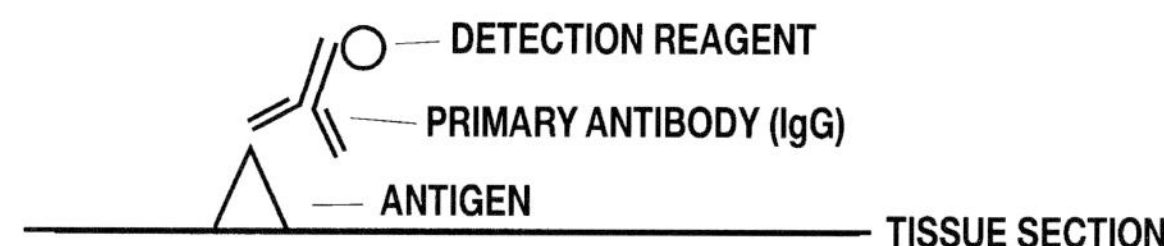

FIGURE 7.1
Direct detection of antigen using tagged primary antibody.

The advantages of the direct detection method are twofold: (1) clean signal with low background, and (2) simplicity. The disadvantages are (1) purification and coupling of the detection reagent to each preparation of primary antibody, and (2) low sensitivity due to lack of amplification. Because of these disadvantages, most investigators use what is called the indirect detection method.

Indirect Method

In this method, the primary antibody is not conjugated with a detection reagent, but rather the detection agent is conjugated to an intermediate that will bind to the primary antibody. These conjugated intermediates can take several forms; we discuss three that are commonly used. One widely used intermediate is the "tagged" secondary antibody (Figure 7.2). In this method, the primary antibody, e.g., rabbit IgG, binds to the antigen. Then to this complex is added a secondary antibody, e.g., goat-anti-rabbit IgG, that binds to the primary antibody. The secondary antibody is also tagged with some detection reagent so that the complex of antigen, primary antibody, and secondary antibody can be visualized. The general method for the indirect detection of antigen involves the following basic steps:

1. Rinse slide containing tissue section with buffer.
2. Remove excess liquid from around section.
3. Incubate with blocking reagents (see Section II.D.5).
4. Apply appropriately diluted primary antibody (see Section II.D.3). Incubate for minutes to hours.
5. Remove antibody; repeat steps 1 and 2.
6. Apply appropriately diluted secondary antibody (see Section II.D.3).
7. Repeat steps 1 and 2.
8. Activate detection reagent (see Section II.D.2).
9. Rinse, counterstain, and coverslip.

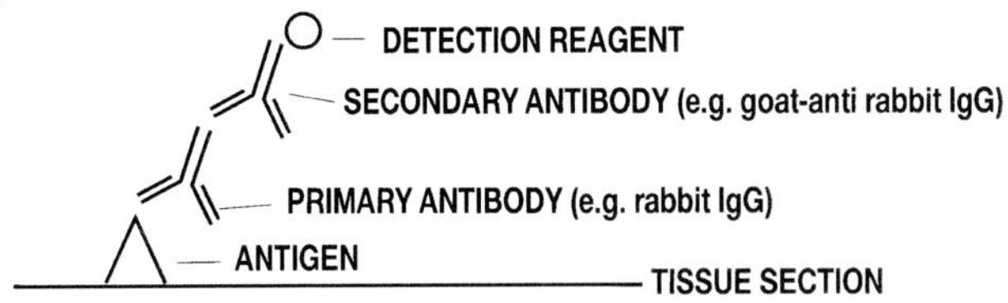

FIGURE 7.2
Indirect detection of antigen using tagged secondary antibody.

The main advantage of this method is that one tagged secondary antibody can be used to visualize a variety of primary antibodies from the same species. This method is also potentially more sensitive than the direct method because several secondary antibodies can bind to different epitopes on the primary antibody. Because each secondary antibody is tagged with a detection molecule, amplification of the detection signal is achieved. A specific protocol using the indirect method is presented in Appendix B.

Another indirect method uses Protein A or Protein G as bridging intermediates. Protein A is a 42-kD protein from *Staphylococcus aureus* that contains four potential binding sites for the Fc fragment of the major immunoglobulin classes from a variety of species. Protein G is a 30- to 35-kD cell wall protein from beta-hemolytic streptococci that binds to the Fc fragment of the major immunoglobulin classes from a variety of species (see Figure 7.3). Both Protein A and G are tagged with some detection reagent. The main advantages of these intermediates is their high affinity for antibodies (IgG) and the lack of interference with the antigen-antibody binding reaction.

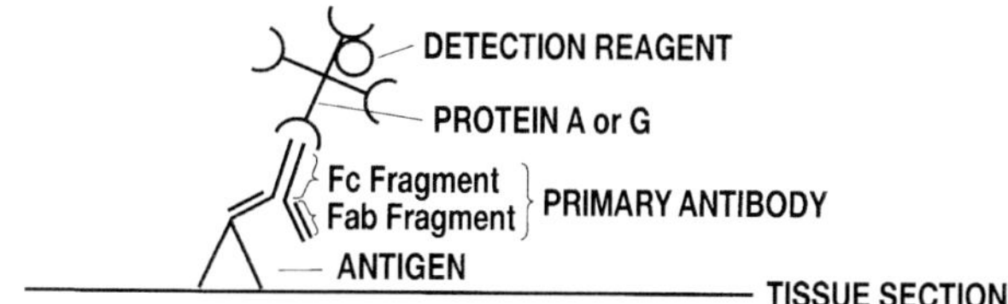

FIGURE 7.3
Indirect detection of antigen using tagged protein A or G.

The final bridging intermediates that we will discuss consist of secondary antibodies that are coupled to biotin. After the biotinylated secondary antibody has complexed with the primary antibody/antigen, the tissue section is then incubated with either avidin or streptavidin, each of which is tagged with some detection reagent (Figure 7.4). Avidin is a 68-kD eggwhite glycoprotein containing four high affinity binding sites for biotin (dissociation constant $> 10^{-15}$). Streptavidin is a 60-kD tetrameric protein isolated from *Streptomyces avidinii* that also contains four high affinity binding sites for biotin. The advantages of using biotinylated secondary antibodies and avidin/streptavidin are the high affintiy of avidin/streptavidin for biotin and the possibility of amplification of the detection signal because of the multivalent nature of avidin/streptavidin. In addition, further amplification can be achieved using preformed avidin/biotin/enzyme complexes (e.g., the ABC kit from Vector Laboratories). These two advantages coupled with low background (particularly streptavidin) combine to make this a sensitive method of detecting antigens.

2. Detection Reagents

Whether one uses the direct or indirect method of detecting antigens in tissue sections, one is faced with the problem of visualizing the antigen/antibody complex. In this section we outline the two major detection reagents, i.e., antibodies coupled with fluorochromes that can be excited by different wavelengths of light and antibodies coupled with various enzymes capable of converting a colorless substrate to a colored product. In addition, we mention

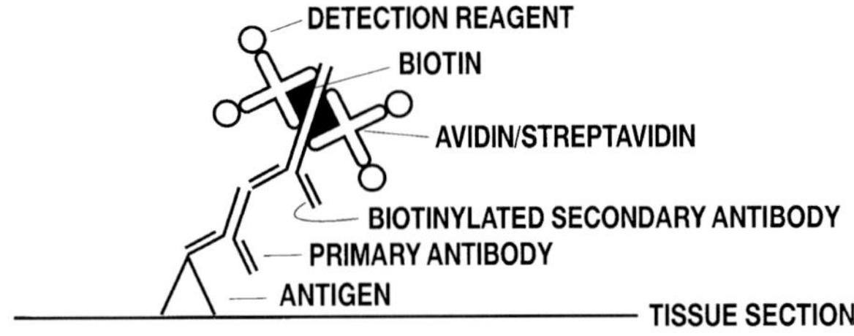

FIGURE 7.4
Indirect detection of antigen using biotinylated secondary antibody and avidin/streptavidin.

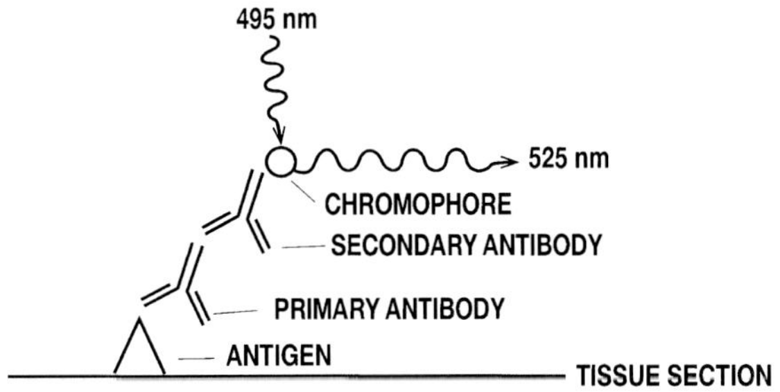

FIGURE 7.5
Detection of antigen/antibody complex using fluorescent-conjugated secondary antibody.

briefly the possibility of using radiolabeled secondary antibodies and autoradiography to visualize specific antigens.

Fluorochromes. Antibodies coupled with fluorochromes that can be activated by different wavelengths of light are widely used in immunohistochemistry (Figure 7.5). Some of the common fluorochromes and their major characteristics are listed in the table below.

Fluorochrome	Excitation	Emission	Color
Fluorescein	495 nm	525 nm	Green
Texas red	596 nm	620 nm	Red
Rhodamine	552 nm	570 nm	Red
Phycoerythrin	488 nm	578 nm	Orange-red
Cy3	552nm	565 nm	Red

Fluorochromes can be conjugated to anti-immunoglobulin antibodies, Protein A/G, or avidin/streptavidin. The advantages of using fluorochromes as detection reagents are (1) high resolution, and (2) possibility of visualizing two (or more) antigens simultaneously by using different fluorochrome-conjugated antibodies. The disadvantages of using flurochromes are (1) required use of an expensive fluorescent microscope, (2) impermanence of fluorescent stain due to photobleaching, and (3) incompatibility with some histological stains.

Plate 1* shows a sagittal section through a day 11 rat embryo immunohistochemically stained for aldehyde dehydrogenase (ALDH). Tissue sections were incubated with a primary antibody (rabbit) directed against ALDH, which was followed by a biotinylated secondary (donkey-anti-rabbit IgG) and incubation with Texas red-conjugated streptavidin.[11] Major sites of ALDH expression are tissues of the developing gastrointestinal tract and kidney.

Enzymes. The most popular method for visualizing antigen/antibody complexes involves the use of antibodies (usually secondary) conjugated with enzymes capable of converting a colorless chromogenic substrate to a colored product (Figure 7.6). The advantages of using enzyme-conjugated rather than fluorochrome-conjugated antibodies are (1) high sensitivity, (2) compatibility with many histological counterstains, and (3) permanence. Disadvantages include (1) low resolution, (2) the problem of endogenous enzyme activity, and (3) difficulty of double staining.

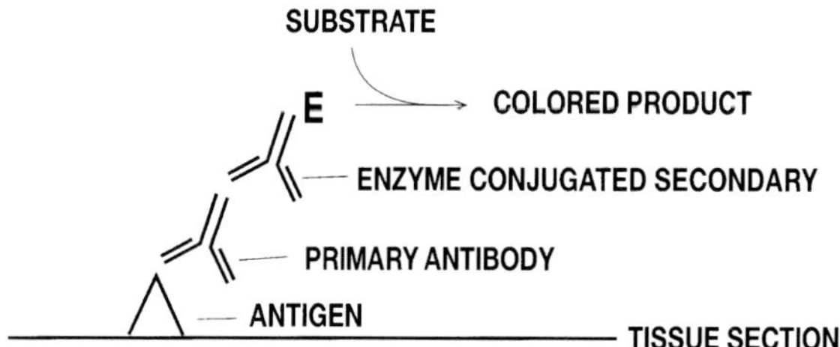

FIGURE 7.6
Detection of antigen/antibody complex using enzyme-conjugated secondary antibody.

One commonly used enzyme is horseradish peroxidase (HRP), which catalyzes the following reaction:

$$HRP + H_2O_2 + \text{electron donor (chromagen)} \rightarrow \text{colored product} + H_2O.$$

Some of the commonly used chromagenic substrates for this enzyme are listed below.

Chromagen	Alcohol Solubility	Product Color
3,3 Diaminobenzedine tetrahydrochloride (DAB)	Insoluble	Brown
DAB + nickel chloride	Insoluble	Black
4-Chloro-1-napthol (CN)	Soluble	Blue
3-Amino-9-ethyl carbazole (AEC)	Soluble	Red
p-Phenylenediamine dihydrochloride/pyrocatechol	Insoluble	Blue/black

* Plate 1 follows page 176.

An example of the use of an HRP conjugated antibody and the DAB substrate is presented in Plate 2.* The tissue section shown in this figure was incubated with a primary antibody directed against a conserved epitope of the small heat shock protein, Hsp 27 (see Appendix B for detailed protocol). The brown DAB staining indicates that Hsp 27 is localized primarily to the heart of the day 11 rat embryo.

A second commonly used enzyme that is conjugated to antibodies is alkaline phosphatase (AP), an enzyme that catalyzes the following reaction:

$$AP + \text{substrate (chromagen)} \rightarrow \text{colored product} + PO_4$$

Commonly used chromagenic subtrates for this enzyme are listed in below.

Chromagen	Alcohol Solubility	Product Color
Napthol-AS-BI-PO_4/New Fuchsin (NABT/NPT)	Insoluble	Red
Bromochloroindoyl PO_4 nitroblue tetrazolium (BCIP/NBT)	Insoluble	Black/purple
4-Chloro-2-methyl benzene diazonium/3-hydroxy-2-naphthoic acid 2,4,-dimethylanilide PO_4	Soluble	Red

An example of the use of a AP conjugated antibody is provided in Plate 3.** The section shown (day 15 rat embryo limb bud) was incubated with a primary antibody directed against clusterin and then with an secondary antibody conjugated with alkaline phosphatase (AP). This complex was then visualized by incubating the tissue section with the AP substrate, BCIP/NBT, the reaction product of which gives a blue-purple color. Distribution of colored reaction product indicates that clusterin is expressed in discrete areas within the digital and interdigital mesenchyme.[10]

Radioisotopes. Radiolabeled secondary antibodies (anti-IgG) are available from Amersham, and the S^{35}-labeled antibodies could be used in conjunction with autoradiography to localize proteins of interest. Using this approach, however, will introduce the complications associated with autoradiography, such as dipping slides in emulsion (in the dark), potentially long exposure times, and having to deal with radioactive substances. One of the advantages of this approach could be sensitivity; however, we are not aware of any studies that compare the sensitivity of radiolabeled vs. enzyme or fluorochrome-conjugated antibodies.

* Plate 2 follows page 176.

** Plate 3 follows page 176.

3. Antibody Dilutions

One of the challenges of immunohistochemistry is to maximize antigen/antibody staining while minimizing nonspecific or background staining, i.e., maximizing the signal-to-noise ratio. Although a number of factors contribute to this ratio, one of the most critical is the optimal antibody dilution. In the next section, we discuss additional factors that contribute to background staining.

Determining the optimal antibody dilution is relatively straightforward, but it cannot be overly emphasized that the systematic determination of this dilution is essential for optimal staining. Moreover, the determination of antibody dilution must be repeated for each new antibody and each new batch of antibody. Regardless of whether one purchases antibodies from a commercial company, receives antibodies prepared by other researchers, or prepares one's own antibodies, the first step in using these reagents is to determine the optimal dilution. Commercial companies usually provide suggested dilutions; however, these should be taken only as guidelines. The same is true for antibodies obtained from other researchers. To determine the optimal antibody dilution, we use the checkerboard titration approach outlined in Figure 7.7.

		1° Antibody		
		1:100	1:1000	1:10,000
2° Antibody	1:100	very good / slight	good / slight	pale / none
	1:1000	good / none	pale / none	pale / none
	1:10,000	pale / none	pale / none	none / none

FIGURE 7.7
Checkerboard titration scheme for determining optimum antibody dilutions.

This example applies to the indirect method of detection involving a primary and a secondary antibody and assumes that little, if any, information is available concerning antibody concentration, avidity, or titer. In this case, three dilutions each differing by a factor of 10 are used in the initial optimization. For each cell in the checkerboard, an assessment is made of the quality of specific staining (numerator in Figure 7.7) and the level of nonspecific or background staining (denominator in Figure 7.7). Based upon the evaluation presented in Figure 7.7, one would conclude that a 1:100 dilution of the primary and 1:1000 dilution of the secondary antibody are optimal. Given that antibodies are expensive to purchase or may be limited in quantity, one may want to establish a secondary titration to further refine the optimal dilutions of both antibodies. For example, a titration would be performed using dilutions differing by a factor of two. If, on the other hand, information is already available concerning a recommended dilution, a titration should be set up in which the recommended dilution is bracketed by two dilutions each differing by a factor of two.

Before we move on to other factors that contribute to the signal-to-noise ratio, we will make a few comments concerning two other parameters that affect antibody binding, i.e., length of incubation with antibody and temperature of incubation. Length of incubation can be as short as several minutes to as long as 48 hours. Again, the optimal length of incubation must be determined empirically for each antibody and is influenced by antibody concentration, antibody affinity, and the need to minimize nonspecific staining. Typically, one of three temperatures is used in immunohistochemistry, and the choice of a particular temperature is dictated by antibody affinity, the antibody dilution, and convenience, i.e., it is often convenient to incubate tissue with primary antibody overnight. If incubations are performed overnight or longer, the temperature is maintained at either 4°C or at room temperature (typically around 20°C). If shorter incubations are used, the temperature is maintained at 37°C to take advantage of the fact that the time to reach equilibrium in antigen-antibody reactions is temperature dependent.

4. Controls

As described in the previous sections, immunohistochemistry involves visualizing the antigen of interest in a tissue section by virtue of a staining reaction. The goal is to show that the observed staining pattern accurately reflects the tissue distribution of the protein of interest; however, before such a conclusion can be reached, a variety of controls must be employed to demonstrate that the staining observed is the result of antibody binding to the protein of interest. Controls are of two major types, those that demonstrate methods specificity and those that demonstrate antibody specificity. Controls for method specificity will show whether or not the observed staining results exclusively from the interaction of primary antibody and corresponding tissue antigen. Controls for antibody specificity will show whether the primary antibody is specific for the antigen of interest. The following two sections highlight the major controls necessary to demonstrate methods and antibody specificities. For an in-depth discussion of immunohistochemical controls, the reader is referred to Reference 2.

Methods Specificity. Several controls must be employed to prove method specificity. For most investigators, the first method of specificity control is to delete the primary antibody from the immunohistochemical protocol and replace it with antibody diluent. In most cases this will result in the complete absence of staining. If staining remains, then such staining is nonspecific (see Section II.D.5 for a discussion of the common sources of nonspecific staining). Larsson[2] recommends a more systematic approach to method specificity controls. For example, if one is using a staining procedure employing primary antibody, biotinylated secondary antibody, and streptavidin peroxidase, Larsson would recommend first deleting all intermediates (primary, secondary antibodies and streptavidin peroxidase). By using this systematic approach, one can most often identify the likely source of any nonspecific staining. For example, if nonspecific staining is observed when the primary or the secondary antibody is deleted but not when the strepavidin peroxidase is deleted, one should suspect endogenous peroxidase activity. A second method of specificity control is to substitute normal serum or preimmune serum from the same source in which the primary antibody was raised. Again, in most instances such controls will result in the complete absence of staining. If staining persists, this would suggest that staining is related either to nonspecific binding of serum IgG to tissue proteins or to endogenous enzyme activity. Both types of nonspecific staining usually can be eliminated with the use of blocking agents (see Section II.D.5). A third method is to assess the extent of staining as a function of different antibody dilutions. Specific staining, i.e., involving antigen and antibody, will decrease and be extinguished as the antibody dilution is increased. Nonspecific staining (for example, endogenous enzyme activity) will not be affected by antibody dilution.

Antibody Specificity. Whereas method specificity is relatively easy to demonstrate, antibody specificity is far more difficult to prove. The most commonly used antibody specificity control is to pre-adsorb the antibody with the antigen used to elicit the antibody. This pre-adsorbed antibody is then substituted for the original primary antibody. In this approach, all antigen binding sites are complexed with antigen; therefore, no interaction with tissue antigen is possible, and staining should be eliminated. This approach works best with antibodies induced to synthetic peptides where sufficient quantities of the antigen necessary for adsorption are available. When using peptides as antigens, it is also necessary to confirm that the resulting antiserum does not contain antibodies to the carrier molecule to which the peptide is complexed. This can be addressed by adsorbing the antibody preparation with the carrier molecule.

If staining persists after adsorption of the primary antibody, this could indicate that (1) the antibody preparation contains other antibodies to contaminating antigens that are also present in the tissue of interest, or (2) the antibody of interest cross-reacts with closely related epitopes on related or even unrelated proteins. To eliminate staining due to the former, one can resort to affinity purification of the antibody of interest, thereby eliminating antibodies to contaminating antigens. Affinity purification is performed by passing the antiserum over a column to which the antigen of interest is bound. Alternatively, one can attempt to eliminate unwanted staining by antibody dilution. Often, contaminating antibodies are present in low concentrations and are sometimes of low affinity compared to the antibody of interest; therefore, it is often possible to dilute the primary antibody preparation to a level that eliminates unwanted, but retains specific, staining.

The most difficult problem to deal with is the binding of the antibody of interest to epitopes similar to the immunizing epitope present on related or even unrelated proteins. There is no straightforward solution to this problem; however, several approaches can be used. First, use Western blotting to confirm the presence of a cross-reacting antigen. If the primary antibody (polyclonal) is cross-reacting with another protein, it is sometimes possible to eliminate "cross-reacting" antibodies by adsorption with the cross-reacting protein (assuming its identity is known). Second, prepare several different antisera from different animals or by different methods in an attempt to find an antibody that lacks nonspecific cross-reactivity. Failing this, the only recourse is to attempt to correlate the observed immunohistochemistry with a biochemical characterization of the antigen.

5. Blocking Reagents

Under ideal circumstances, the fluorescent or colored signal observed in immunostained tissue sections will exactly match the spatial/temporal distribution of the protein of interest. Frequently, however, this ideal is not achieved initially, and even in tissues in which the distribution of the protein of interest is known, it is not unusual to observe less intense staining in all or most tissues in addition to the expected staining. This so-called nonspecific (or background) staining is one of most frequently encountered problems in immunohistochemistry. In this section, we discuss the major sources of this problem along with suggested solutions.

Antibody Binding. Ideally, the primary antibody will bind only to the epitope used to elicit the antibody and, if the indirect detection method is used, the secondary antibody will bind only to the primary antibody. This is the ideal, which is rarely realized in practice because besides being antigen-binding molecules, antibodies are also proteins and therefore subject to various nonantigen protein-protein interactions. Two of these interactions that may impact the immunohistochemical process are hydrophobic and ionic interactions between antibodies and nontarget proteins.

Protein hydrophobicity, the tendency of neutral amino acids in proteins to repel water, not only confers stability to the tertiary structure of proteins, but also facilitates hydrophobic interactions between proteins. Thus, antibodies have a tendency to participate in nonspecific binding through hydrophobic interactions. This tendency is further enhanced by aggregation due to storage, polymerization, conjugation to other proteins, and by fixing tissues with cross-linking fixatives. In practice, these unwanted hydrophobic interactions can be minimized by adjusting the ionic strength of antibody buffer, adding a detergent (e.g., Tween-20, Triton X-100), or raising the pH of the antibody buffer. Perhaps more effective is blocking nonspecific protein-protein interactions by incubating tissue sections with protein solutions such as BSA (3%), nonfat dry milk (4%), or serum from the species used to produce the secondary antibody prior to incubation with the primary antibody. We routinely add the appropriate blocking agent to the primary and secondary antibody preparations.

Ionic interactions between antibodies and nontarget proteins can occur because antibodies typically carry a net positive charge over the pH range (7.0 to 7.8) normally used in antibody buffers. This positive charge facilitates ionic interactions with negatively charged molecules in the tissue. These interactions can be minimized by raising the ionic strength of the antibody buffer by elevating the salt concentration (0.1 to 0.5 *M* NaCl). In addition, such interactions are minimized by blocking with the protein solutions described previously.

Enzyme Activity. If one uses enzyme-conjugated antibodies to visualize antigens, another source of background staining may be encountered, i.e., endogenous enzyme activity. For example, endogenous peroxidases (e.g., catalase) as well as "pseudoperoxidases" (hemoproteins such as hemoglobin, myoglobin, and cytochromes) may present background problems in some tissues. Endogenous peroxidase activity is inactivated by incubating the tissue with 3% H_2O_2 for 8 to 10 minutes or methanolic-H_2O_2 (1 part 3% H_2O_2:4 parts absolute methanol) for 20 minutes prior to incubation with the primary antibody. For phosphatase-conjugated antibodies, one must be concerned with endogenous phosphatase activity. For most tissues, endogenous phosphatase activity can be blocked effectively by incubating tissue sections with 0.1 m*M* levamisol (Sigma). The intestinal alkaline phosphatase conjugated to antibodies is not blocked by levamisol. Tissues containing intestinal-type alkaline phosphatase cannot be effectively blocked by levamisol. In this instance, an alternative solution is to use a fluorochrome conjugated antibody.

Avidin/Biotin. Similarly, if one uses biotinylated antibodies and avidin/streptavidin conjugated to enzymes as the detection method, another potential source of background is endogenous biotin, a vitamin found in many tissues and a coenzyme bound to a variety of enzymes and proteins. To block endogenous biotin, incubate tissue sections with 0.1% avidin for 20 minutes to bind available biotin and then block all avidin sites by incubating tissue sections with 0.01% biotin for 20 minutes prior to incubating with the primary antibody. In addition, avidin contains carbohydrates that may bind to lectin-like substances. This binding can be blocked by incubating tissue sections with 0.1 *M* alpha-methyl-D-mannoside. The alternatives are to use streptavidin in place of avidin because the former does not contain carbohydrate moieties or Avidin D (Vector Laboratories), for example, that has been selected for low nonspecific binding.

6. Unmasking Antigenic Sites

In some instances, even in tissues known to contain the antigen of interest (for example, based upon Western blotting) one is faced with a complete lack of staining. Assuming there is no trivial explanation for the lack of staining, such a result usually indicates that the protein epitope of interest is masked. This problem is particularly prevalent when using tissues that have been fixed by cross-linking fixatives. Although a variety of "unmasking" methods (Appendix C) has been used, the most widely effective method involves using various proteases (e.g., trypsin, pepsin, chymotrypsin, and pronase) to unmask the antigenic epitope of interest. A more recent method involves microwaving tissue sections to unmask antigenic epitopes. If these approaches prove unsatisfactory, an alternative approach is to turn to frozen tissue and cryostat sections, where one can minimize the extent of fixation.

E. Monoclonal vs. Polyclonal Antibodies

For many antigens, one has a choice of using either polyclonal or monoclonal antibodies and this section highlights some of the advantages and disadvantages of each. Polyclonal antibodies are raised by injecting "purified" proteins into an animal, most often a rabbit. Because different antibody-producing cells can produce antibodies against many different regions (epitopes) of the injected protein, many clones of antibody-producing cells each directed against a particular epitope are produced, hence the term polyclonal. Monoclonal antibodies are produced by injecting a protein into a mouse. At the appropriate time the mouse is killed and B cells are removed from the spleen and then fused with mouse myeloma cells, producing a hybrid cell or hybridoma. These hybridoma cells have the ability to produce antibody (B cell contribution) and can be maintained in culture essentially indefinitely (myeloma contribution). From a starter culture of such hybridoma cells, clones of plasma cells each producing antibodies to one specific epitope (hence the term monoclonal) are selected.

A quick review of journals publishing immunohistochemical data will reveal that both polyclonal and monoclonal antibodies are widely used. Furthermore, there is no universal advantage that can be attributed to either type of antibody. In most instances, the choice of monoclonal or polyclonal is one of personal preference. Nonetheless, some of the advantages of polyclonal vs. monoclonal antibodies are listed below.

Polyclonal Antibodies

1. Higher affinity (generally)
2. Wider reactivity
3. Greater stability at varying pH and [salt]
4. Less costly and simpler to produce

Monoclonal Antibodies

1. Virtually unlimited supply
2. Narrow specificity
3. High homogeneity
4. Antibodies to rare antigens feasible

F. Immunohistochemistry and Image Analysis

Assuming that the protocols outlined in Sections II.2 and II.3 were followed, one should have an initial assessment of where the protein of interest is expressed in the tissue section(s). In many instances, this may be the only information that is sought; e.g., does the prosencephalic neuroepithelium of a day 9 mouse embryo express protein X, or where in the day 9 mouse embryo is protein X expressed? In other instances, however, additional information is sought, e.g., the number of prosencephalic neuroepithelial cells expressing protein X or the relative abundance of protein X in prosencephalic vs. rhombencepahlic neuroepithelia. Obtaining this kind of information manually is at best tedious and at worst impossible. Recently, however, a number of commercial companies (Appendix D) have begun to offer hardware and software to capture and analyze immunohistochemical images. Because image analysis needs will be dictated by the individual investigator's research, it is not possible within the framework of this chapter to provide even a generic discussion of image analysis. Instead, the reader is referred to the excellent discussion of techniques for image analysis by Read and Rhodes in Reference 7. In addition, we suggest that investigators discuss their specific needs with representatives of commercial companies that market products for image analysis. This is often the quickest way to determine whether commercially available products will meet your specific needs.

G. Whole-Mount Immunohistochemistry

The preceding sections have outlined methods for visualizing proteins in tissue sections. These methods are excellent for the two-dimensional analysis of protein expression patterns in the mammalian embryo; however, these methods provide little information concerning three-dimensional expression patterns. Fortunately, methods have recently been developed that allow proteins to be localized in unsectioned tissues/embryos using specific antibodies, i.e., whole-mount immunohistochemistry. In Appendix E, we present a protocol that we have used to localize the engrailed protein in early postimplantation rat embryos in Appendix E. For additional information concerning whole-mount immunohistochemistry, the reader is referred to Reference 9. Plate 4* shows a photomicrograph of a day 11 embryo immunohistochemically stained to show engrailed protein expression. This photomicrograph clearly reveals the expression pattern of this protein in the context of the entire three-dimensional embryo, an expression pattern limited to the caudal end of the mesencephalon and the rostral portion of the rhombencephalon.[8] To gain a more detailed examination of protein expression, stained embryos can be embedded in paraffin and sectioned.

---

* Plate 4 follows page 176.

III. Summary

In this chapter, we have highlighted (1) the major points that must be considered before undertaking the immunohistochemical localization of a protein, (2) generic and specific protocols for localizing proteins, and (3) examples from our own work demonstrating the localization of several different proteins in the early postimplantation mammalian embryo. Immunohistochemistry is a powerful tool that can reveal spatial patterns of protein expression, information that cannot be obtained using other protein detection methods. Not only does immunohistochemistry reveal where a particular protein is expressed, this approach also produces artistically striking images that are often breathtaking. Despite the power and artistry of this method, the ultimate value of the images obtained is directly related to the quality of the controls that are performed. "Garbage in, garbage out" is as true for immunohistochemistry as it is for computers.

References

1. Harlowe, E. and Lane, D., Eds., *Antibodies: A Laboratory Manual,* Cold Spring Harbor Laboratory Press, Cold Spring Harbor, NY, 1988.
2. Larsson, L.-I., *Immunocytochemistry: Theory and Practice,* CRC Press, Boca Raton, FL, 1988.
3. Beltz, B.S. and Burd, G.D., *Immunocytochemical Techniques: Principle and Practice,* Blackwell Scientific Publications, Cambridge, MA, 1989.
4. Polak, J.M. and Noorden, S.V., *Immunocytochemistry: Modern Methods and Applications,* 2nd ed., John Wright & Sons, Bristol, 1986.
5. Manson, M.M., Ed., *Methods in Molecular Biology 10: Immunochemical Protocols,* Humana Press, Totowa, NJ, 1992.
6. Cuello, A.C., Ed., *Immunohistochemistry II,* John Wiley & Sons, New York, 1993.
7. Beesley, J.E., Ed., *Immunocytochemistry: A Practical Approach,* IRL Press, New York, 1993.
8. Cunningham, M.L., MacAuley, A., and Mirkes, P.E., From gastrulation to neurulation: transition in retinoic acid sensitivity identifies distinct stages of neural patterning in the rat, *Dev. Dynamics,* 200, 227-241, 1994.
9. Davis, C.L., Whole-mount immunohistochemistry, *Methods Enzymol.,* 225, 502-516, 1993.
10. Little, S.A. and Mirkes, P.E., Clusterin expression during programmed and teratogen-induced cell death in the postimplantation rat embryo, *Teratology,* 52:41-54, 1995.
11. Mirkes, P.E., Ellison, A., and Little, S.A., Resistance of rat embryonic heart cells to the cytotoxic effects of cyclophosphamide does not involve aldehyde dehydrogenase-mediated metabolism, *Teratology,* 43, 307-318, 1991.

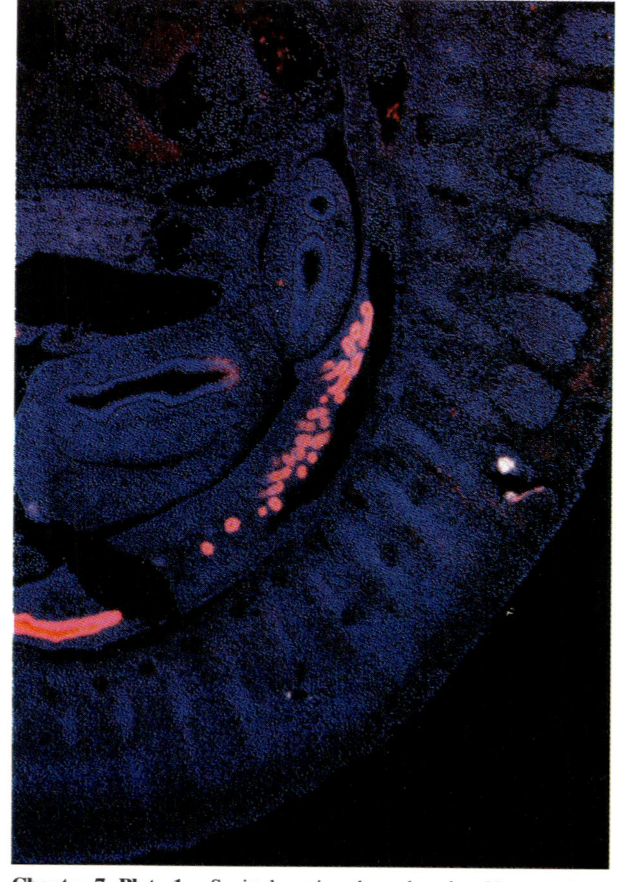

Chapter 7, Plate 1. Sagittal section through a day 11 rat embryo immunohistochemically stained for aldehyde dehydrogenase.

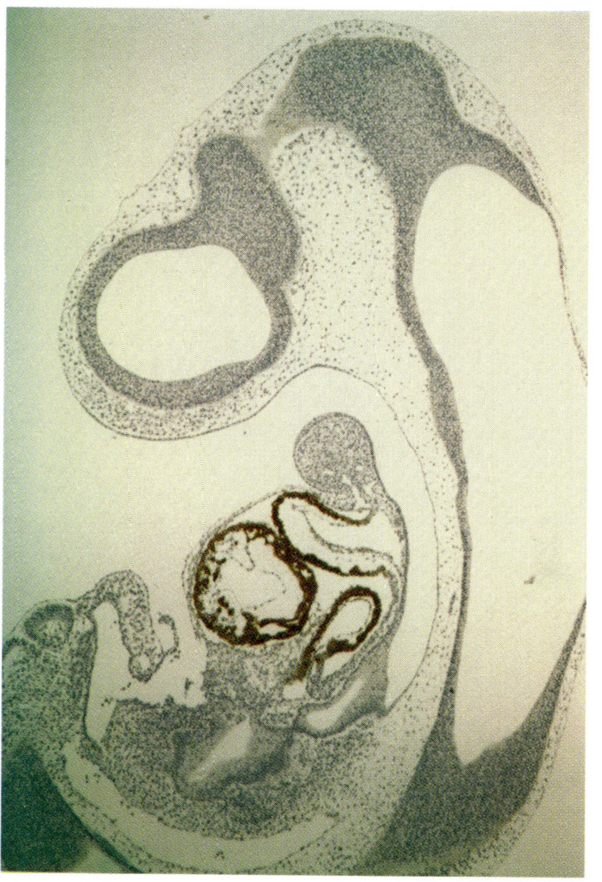

Chapter 7, Plate 2. Sagittal section through a day 11 rat embryo immunohistochemically stained for heat shock protein 27 (Hsp 27).

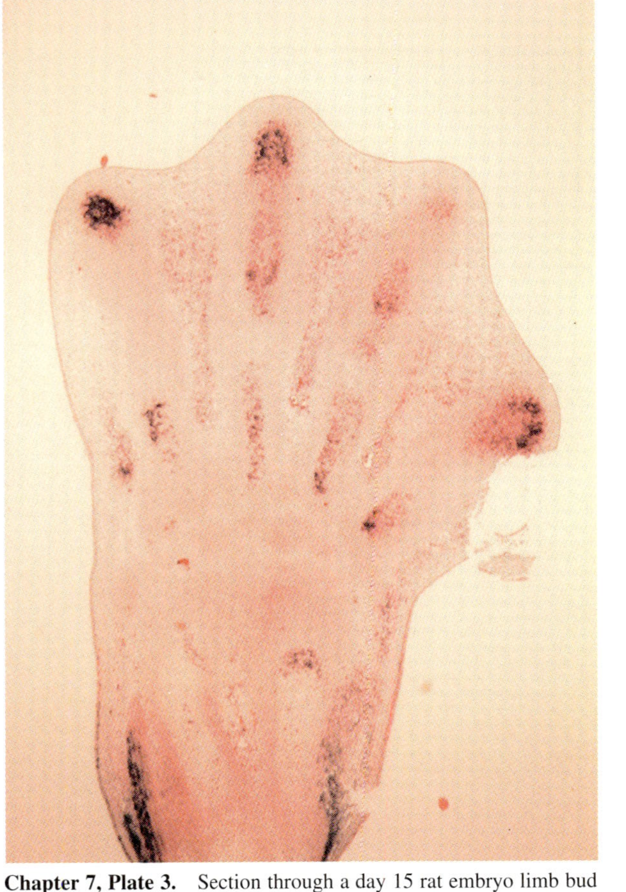

Chapter 7, Plate 3. Section through a day 15 rat embryo limb bud immunohistochemically stained for clusterin.

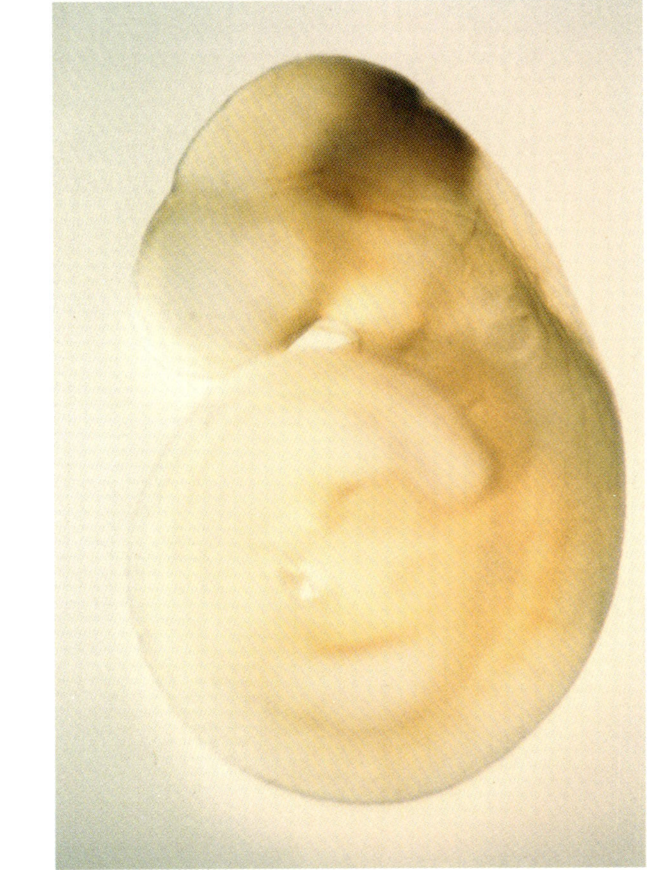

Chapter 7, Plate 4. *Engrailed* expression in a day 11 rat embryo detected by whole-mount immunohistochemistry.

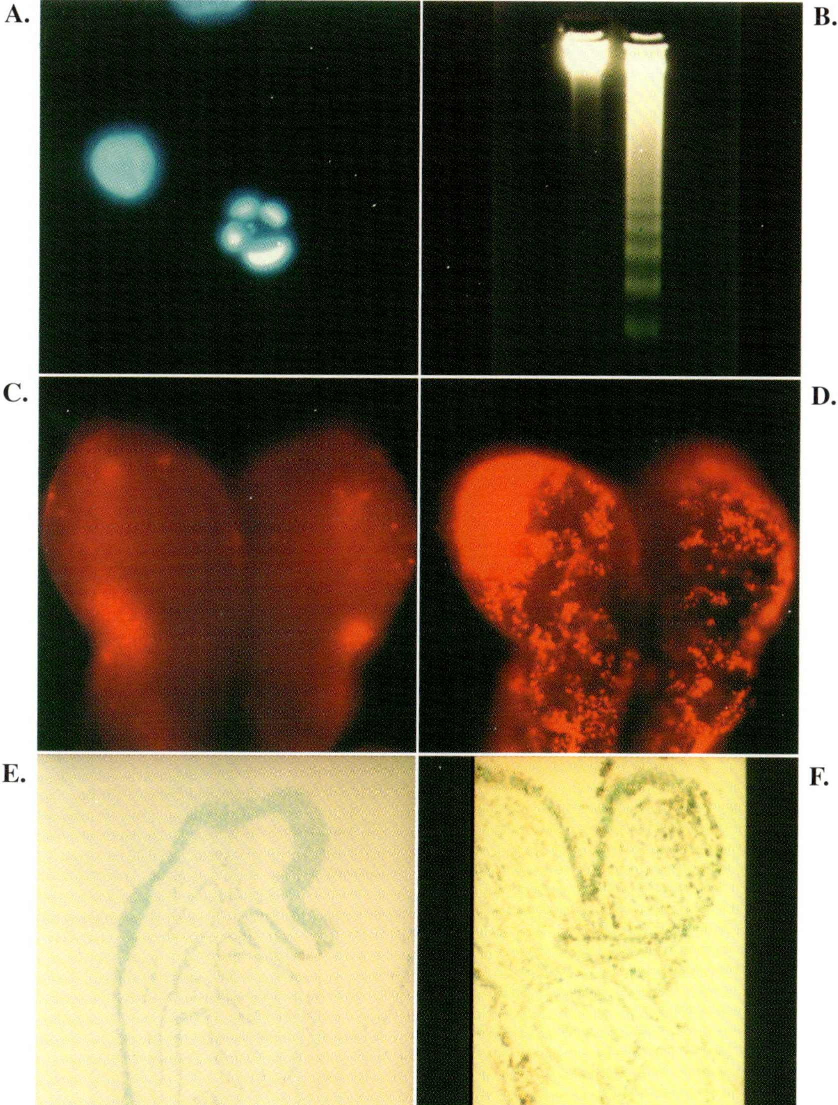

Chapter 8, Plate 1. Cellular techniques for teratogen-induced apoptosis. **(A)** Nuclear changes in apoptotic cells. Leukemia cell induced to apoptosis with 5 μ*M* 2-chlorodeoxyadenosine (2-CdA) in culture and stained with DAPI (see Referene 12). **(B)** Classical nucleosomal ladder in day 9 mouse embryos. Embryos were cultured for 3 h in the presence of 1 m*M* adenosine (left lane) or 0.1 m*M* 2′-deoxyadenosine and pentostatin (right lane); DNA was monitored by the endonucleotlytic assay (see Reference 17). **(C,D)** AO-staining: Fluorescence micrograph of the head-fold of (C) normal day 8 mouse embryo and (D) embryo intoxicated with 2′-deoxyadenosine displaying increased AO-positive staining, photographed under rhodamine optics (see Reference 17). **(E,F)** TUNEL-assay: Light micrograph of (E) normal day 8 mouse embryo section and (F) embryo 4 h after teratogenic treatment with 2-CdA, counterstained with methyl green; TUNEL-positive nuclei cell death was induced in the neuroepithelium but not the primitive heart (see Reference 37).

Appendix A: Suppliers of Antibodies or Immunohistochemical Reagents

Supplier	Location
Accurate Chemical & Scientific Corp.	San Diego, CA
Affinity BioReagents, Inc.	Neshanic Station, NJ
American Qualex	La Mirada, CA
Amersham/Life Science	Arlington Heights, IL
Becton Dickinson	Mountain View, CA
Boehringer Mannheim	Indianapolis, IN
CalBioChem	San Diego, CA
CalTag	San Francisco, CA
Cappel/Organon Technika	Durham, NC
Cytokine Research Products/Genzyme Corp.	Boston, MA
Dako Corp.	Carpinteria, CA
E & Y Laboratories, Inc.	San Mateo, CA
Gibco/BRL	Grand Island, NY
ICN	Irving, CA
Kirkegaard and Perry Laboratories	Gathersburg, MD
Novocastra Labs, Ltd. (Vector Distrib.)	Burlingame, CA
Oncogene Science, Inc.	Uniondale, NY
PharMingen Research Products	San Diego, CA
Pierce	Rockford, IL
R & D Systems	Minneapolis, MN
Santa Cruz Biotechnology, Inc.	Santa Cruz, CA
Sigma Chemical Co.	St. Louis, MO
StressGen Biotechnologies Corp.	Victoria, BC, Canada
The Binding Site, Inc.	San Diego, CA
UBI (Upstate Biotechnology, Inc.)	Lake Placid, NY
Vector	Burlingame, CA

A valuable resource is the *Linscott Directory*; 4877 Grange Rd.; Santa Rosa, CA (707-544-9555), which lists antibodies and their sources. Another valuable reference is the journal *Biotechnology,* 13(12), 1264-1270, 1995, which lists commercial suppliers of antibodies.

Appendix B: Protocol for Immunohistochemical Staining Using the Indirect Method

Fix postimplantation mouse/rat embryos in Carnoy's fixative. Embed in paraffin and prepare 5-micron sections.

Procedure

1. Deparaffinize slides: 58°C for 30 minutes.

   Histoclear: 5 minutes × 2

   100% ethanol: 2 to 5 minutes × 3
2. 0.3% H_2O_2 in methanol for 20 minutes.
3. 70% methanol for 2 minutes.
4. PBS for 2 minutes × 2.
5. 1% sheep serum (species of secondary Ab) for 1 hour.
6. Rinse briefly in PBS.*
7. Overlay sections with primary Ab (mouse-anti-Hsp27) diluted 1:1000 in 1% sheep serum/PBS.
8. Incubate overnight in humidifying chamber at room temperature.
9. Flush slide with PBS.
10. Wash in 0.1% Tweens in PBS for 5 minutes.
11. Wash in PBS, 5 minutes × 3.*
12. Overlay with secondary Ab (biotinylated sheep-anti-mouse) diluted 1:200 in PBS.
13. Incubate for 1 hour at room temperature.
14. Repeat steps 9 to 11.
15. Overlay sections with avidin-horseradish peroxidase complex (Vector Labs; ABC reagent prepared according to vendor instructions).
16. Incubate for 30 minutes at room temperature.
17. Repeat steps 9 and 11 (eliminate step 10).
18. Apply freshly prepared DAB solution (Sigma Fast DAB tablets).
19. Allow color reaction to develop 12 to 15 minutes. (DAB is carcinogenic; wear gloves. Drain used DAB into hazardous waste container.)
20. Wash slides in H_2O for 5 minutes × 3.
21. Counterstain briefly (5 seconds) in hematoxylin (or methyl green for 1 minute) followed with appropriate washes and dehydration.
22. Coverslip using Histomount.

---

* When overlaying sections with Ab or reaction solutions, dry the back of the slide and carefully around the section. Do not allow the section to dry out. Place slide in a humidifying chamber and immediately and carefully overlay appropriate solution without disturbing the section.

Appendix C: Protocols for Unmasking of Antigens

Protease Digestion

1. Pepsin: 2 to10 μg/ml pepsin in 0.01 *N* HCl (pH 2.5), 5 to 30 minutes at room temperature (Santa Cruz; SIGMA) or 1 mg/ml in 0.5 *M* HAc for 1 to 2 hours at 37°C.
2. Trypsin: 0.1% trypsin in 20 m*M* Tris (pH 7.8) with 0.1%$CaCl_2$; expose sections 2 to 20 minutes at room temperature.
3. Pronase E: 0.05% in PBS for 4 to 8 minutes. at 37°C.
4. Target unmasking fluid (TUF) is available from Pharmingen cat. no. 70001T, also from Boehringer/Manheim, cat. no. 1 666 363.

Microwave Oven Heating

Slides with deparaffinized sections are immersed in 0.01 *M* citrate buffer (pH 6.0) contained in a glass or plastic jar fitted with a loose-fitting lid. Place jar in microwave oven and heat 2 × 5 minutes. Solution must come to a boil each time. Sections must never become dry, and H_2O may need to be added between heating cycles if volume is reduced by evaporation. Allow sections to cool at room temperature for 15 to 20 minutes. Also, a commercial unmasking solution plus detailed protocol are available from Vector Labs (cat. no. H-3300).

Appendix D: Selected Suppliers of Microscopy and Image Analysis Equipment

Axon Instruments, Inc.
Bio-Rad Laboratories
Inovision Corp.
Intracellular Imaging, Inc.
Jandel Scientific Software
Molecular Dynamics, Inc.
Nikon, Inc.
Olympus America, Precision Instruments Division
Oncor Imaging
Princeton Instruments, Inc.
Scanalytics/CSPI
Universal Imaging Corp.
Carl Zeiss, Inc.

A useful reference for image analysis software and hardware is *Biotechnology*, 13(12), 1297, 1995.

Appendix E: Protocol for Whole-Mount Immunohistochemistry Engrailed (alpha-hb-1)

Procedure

1. Collect embryos in HBSS.
2. Fix in methanol DMSO (4:1) overnight at 4°C.
3. Add 30% H_2O_2 to results of step 2 for final = (4:1:1); incubate at room temperature for 6 to 8 hours to bleach embryos and block endogenous peroxidases. The embryos can then be stored at –20°C in methanol for a few weeks.
4. Rehydrate embryos in 50% methanol, 15% methanol, and PBS for 30 minutes each. Be careful, as the PBS/methanol mixture gets warm when first mixed; cool on ice prior to rehydration.
5. Incubate twice in PBSMT* for 1 hour each at room temperature.
6. Incubate overnight at 4°C with primary antibody diluted in PBSMT (1:1000 for En-Ab).
7. Wash with PBSMT, twice with cold and three times with room temperature, 45 minutes each.
8. Incubate overnight with secondary antibody at 4°C diluted in PBSMT (1:200 dilution biotinylated donkey anti-rabbit for En-Ab protocol).
9. Repeat step 7.
10. Wash twice briefly with PBT** (large volume) to remove milk, etc.
11. Incubate with streptavidin-peroxidase (1:200 dilution): 1 hour at 37°C, 1 hour at room temperature while rocking.
12. Repeat step 10, × 20 minutes.
13. Incubate embryos in 0.3 mg/ml DAB*** in PBT at room temperature for a minimum of 20 minutes (the color may be enhanced by adding 0.5% $NiCl_2$ to the DAB solution).
14. Add H_2O_2 to the DAB and embryos to equal 0.03% (10 µl/ml of DAB solution) and incubate at room temperature until the color looks good. It can be very FAST!
15. Rinse in PBT and dehydrate in a methanol series: 30, 50, 80, 100%, for 30 minutes each.
16. Embryos can be cleared with benzyl alcohol: benzyl benzoate (1:2).

* PBSMT = 2% milk powder, 0.1% Triton X-100 in PBS.
** PBT = 0.1% Triton X-100 in PBS, 0.2% BSA.
*** DAB = Diaminobenzidine — carcinogenic!

Part III

Cellular Techniques

Chapter

Cellular Techniques for Teratological Cell Death

Thomas B. Knudsen

Contents

0-8493-3342-3/97/$0.00+$.50

I. Introduction

Cell death is an important part of normal development. It mediates the involution of vestigial structures of importance to phylogenetic but not ontogenetic development, reduces superfluous cells in localized territories, terminates transitory cells after these cells have completed their developmental functions, suppresses the propagation of abnormal or otherwise damaged cells which fail to differentiate or do so inappropriately, and minimizes potential damage from harmful cell types.[1] Many drugs and chemicals that induce excessive cell death in the developing embryo also cause structural malformations in the fetus.[2] Furthermore, dysregulation of the expression and expanse of normal programmed cell death zones has been proposed to explain syndromes of malformations which bear no obvious relationship to one another.[3] Thus, monitoring the patterns of cell death in a developing system provides useful information regarding the pathogenesis of abnormal development. This chapter outlines some of the cellular techniques currently used to monitor teratogen-induced cell death in the mammalian embryo.

II. Modes of Cell Death

Two mechanisms account for the death of cells in a broad spectrum of tissues, stages of differentiation, and circumstances of induction.[4,5] One mechanism, apoptosis, is a signal-induced and gene-directed response of cells to physiological stimuli or specific molecular damage.[6,7] The other, necrosis, is a passive cascade primarily related to the lethal impairment of cellular homeostasis.[8,9] Environmental stress conditions which induce cells to necrosis may, under weaker exposures, initiate apoptosis. Weak stresses that do not disrupt cellular homeostasis produce sublethal lesions that are recognized by cellular surveillance systems and tranduced into a signal for apoptosis. As the stress reaches a critical load, the increased damage impairs cellular homeostasis and necrosis presides.[10,11] Perhaps importantly, most teratogen-induced deaths display the morphological features of apoptosis.[2,3]

A. Necrosis

Necrosis or "lethal cell injury" results from severe environmental stress. Necrotic conditions may include high-level toxicant exposure, ischemia, complement attack, or direct cell trauma.[4,5,8-11] Regions of contiguous cells are commonly struck as an infarct. When a cell is induced to necrosis, it undergoes a relatively consistent sequence of ultrastructural changes which are passive in the sense that they occur without the need for cellular energy metabolism. There is clumping of loosely aggregated chromatin beneath the nuclear envelope, dilation of endoplasmic reticulum, partial dispersal of ribosomes, and condensation of the mitochondrial matrix. Eventually the mitochondrial matrix gradually swells and flocculent matrical densities appear. This is characteristic of an injured cell where mitochondrial respiration is no longer intact. Cellular organelles fall apart; disruption of the nucleus (karyolysis) and plasma membrane (cytolysis) results in eosinophilic "ghosting" of a necrotic cell when viewed by light microscopy. Plasma membrane disruption renders the cell permeable to vital dyes such as trypan blue. In the same way, intracellular contents leak into the extracellular milieu and the sudden release of hydrolases, proteases, inflammatory cytokines, and other kinds of intracellular macromolecules initiates a harmful inflammatory reaction to produce secondary scarring of the surrounding tissue; consequently, signs of necrosis may persist for several days.

B. Apoptosis

Apoptosis or "physiological cell death" involves cellular condensation and fragmentation rather than swelling and dissolution.[4-7] This process, originally referred to as "shrinkage necrosis", was distinctively renamed to give recognition to a fundamental role in tissue homeostasis.[6] Apoptosis is the form of cell death most frequently encountered during programmed cell deletion in normal embryonic development, metamorphosis, and teratogenesis.[2] Importantly, it is an energy-dependent process that specifically affects individual cells rather than contiguous groups of cells. The sequence of ultrastructural changes are essentially the same in a wide variety of apoptotic cells. During the early phase, dense masses of chromatin aggregate beneath the nuclear envelope, and the cell undergoes violent development of surface protuberances. The nucleus often breaks into discrete fragments (karyorrhexis) which disperse into the corresponding cellular protuberances. A typical apoptotic nucleus is shown in Plate 1A.*[12] Cellular protuberances of varying size and shape separate from the cell as the plasma membrane is sealed behind them. Consequently, there is no leakage of intracellular contents into the extracellular milieu. The composition of a particular "apoptotic body" depends on the cellular constituents that happen to be present in the cytoplasmic protuberance that gave rise to it. Some consist largely of condensed chromatin, whereas others may contain cytoplasmic elements only. Apoptotic bodies may be extruded into a lumen but more often are engulfed by neighboring tissue cells and macrophages. Herein they undergo a series of intracellular degenerative changes sometimes referred to as "secondary necrosis". Since potentially toxic contents of dying cells are internally digested within phagosomes, a local inflammatory reaction is avoided. This is an important difference between apoptosis and necrosis.

---

* Plate 1 follows page 176.

Low incidences of apoptosis can have measurable impact on tissue mass. This is so because the morphological stages of apoptosis are brief, perhaps only about 3 hours, and in a tissue where as few as 0.5% cells are visibly apoptotic per hour the net cell loss after several days may reach 25%.[13] Even a small amount of teratogen-induced apoptosis in a rapidly growing organ rudiment has the potential to manifest as a major tissue deficiency. Cellular techniques (apart from electron microscopy) which have been applied successfully to the analysis of teratogen-induced apoptosis will be the focus of the remainder of this chapter (Table 1).

TABLE 8.1
Cellular Techniques for Apoptosis Detection

Method	Advantage	Drawback
Supravital staining	General survey of apoptosis	Opacity limits size of specimen
	Rapid and simple	Compromised by fixation
Endonucleolytic assay	Diagnostic of apoptosis	Not all apoptotic cells show ladder
	Can be quantitated by densitometry	Requires large numbers of apoptotic cells
TUNEL labeling	Accurate localization histologically	Must be optimized for specimen
	Specific for free 3′-OH of DNA	Also labels nonapoptotic DNA breaks

III. Supravital Staining

A. Principle

Vital dye exclusion assays such as trypan blue have limited value in apoptosis detection because they are based on loss of selective membrane permeability. An apoptotic cell may be erroneously classified as viable if it excludes trypan blue. Supravital staining techniques demonstrate apoptotic bodies in whole, living embryos and thus provide a general survey of the magnitude, localization, and timing of apoptosis in the specimen. Two of the more popular supravital dyes are Nile blue sulfate[14,15] and acridine orange (AO).[16,17] Both are actively accumulated by apoptotic bodies in an ATP-dependent manner, by virtue of intact organelle structure and metabolic competency, and they bind to the condensed nucleic acids.[18]

B. Protocol

1. AO (Molecular Probes, Eugene OR; Sigma Chemical Co., St. Louis, MO): 10 mg/ml stock solution in 0.1 *N* HCl stored at 4°C in the dark and diluted to 5 mg/ml with Hank's balanced saline solution (HBSS) just before use.

2. After removing extra-embryonic membranes incubate living embryos (or tissue) in AO solution for 5 minutes at 37°C, rinse 1 × 5 minutes in HBSS at room temperature.
3. Transfer embryo to microscope slide chamber in a few drops of HBSS and immediately examine by fluorescence microscopy. The broad spectrum of AO fluorescence can be detected with green (FITC) or red (rhodamine) optics.

C. Results

Acridine orange-positive cells appear bright yellow against a green background (FITC optics) or bright red against a dark red background (rhodamine optics). An example of the latter is shown for day 8 mouse embryos where large-scale apoptosis has been induced by 2′-deoxyadenosine intoxication (Plates 1C and 1D). AO is not excluded from necrotic cells; however, the loss of metabolic potential renders these cells incompetent to concentrate AO, and so AO-staining is specific for apoptosis.[19] Staining is compromised by tissue fixation, so the embryos must be photographed within a few minutes. Occasional discrepancy may occur in tissues that contain high amounts of glycosaminoglycans, which also stain with AO. Also, the quantitative power of supravital staining is limited because apoptotic bodies from one dying cell may be engulfed by several phagocytes, or one phagocytic cell may contain debris from several apoptotic corpses.

IV. Endonucleolytic Assay

A. Principle

In 1980, Wyllie[20] reported that the morphological changes in an apoptotic nucleus were accompanied by the degradation of DNA at regular intervals of 180 to 200 base pairs (bp). When the low molecular weight DNA fraction was analyzed by agarose gel electrophoresis and ethidium bromide staining, a "ladder" was observed. In stark contrast, DNA from necrotic cells was randomly degraded to an electrophoretic smear. The subunit size of the apoptotic DNA ladder corresponded to the length of DNA wrapped around the histone core of a nucleosome (185 bp), and it was concluded that chromatin condensation in apoptosis was accompanied by cleavage of DNA in the linker region between nucleosomes. Although nucleosomal laddering is a common diagnostic test for apoptosis, the characteristic apoptotic nuclear morphology may occur in its absence and *vice versa,* indicating that regular DNA cleavage is neither necessary nor sufficient for apoptosis.[21,22] Nucleosomal laddering is preceded by an obligate degradation of chromatin to larger pieces of DNA 50 and 300 kilobase pairs (kbp) in size.[23-25] These large fragments represent single (50 kbp) and hexameric (300 kbp) loop structures which are released from their attachment sites to the nuclear matrix. Oligonucleosome fragments that form the characteristic nucleosomal ladder appear when the 50 kbp fragments are degraded by an apoptotic endonuclease.

B. Protocol

1. Suspend tissues in 1 ml lysis buffer (5 m*M* Tris-HCl, 20 m*M* EDTA, 0.5% Triton X-100, pH 8.0) containing 0.1 mg/ml proteinase K (autodigested at 37°C for 2 hours) and incubate at 37°C for 6 hours.
2. Extract the digest with an equal volume of phenol-chloroform-isoamyl alcohol (25:24:1, v/v/v) and re-extract with an equal volume of chloroform:isoamyl alcohol (24:1, v/v).
3. Dialyze the upper aqueous phase against buffer (0.01 *M* Tris-HCl, pH 7.5, containing 1 m*M* EDTA) overnight at 4°C.
4. Digest the dialyzed samples with 0.1 μg/ml RNase (DNase-free) for 5 hours at 37°C; repeat dialysis step.
5. Concentrate samples to an approximate volume of 0.2 ml in a Savant Speed-Vac and determine the final DNA concentration by UV absorbance at 260 nm.
6. Load 10 μg DNA per lane on 1.8% agarose gels containing 1 mg/ml ethidium bromide; the electrophoresis buffer consists of 4 m*M* Tris-HCl, pH 8.3, 2 m*M* sodium acetate, 2 m*M* EDTA, and 1 μg/ml ethidium bromide. Size markers may include pBR322-Hinf digest or λDNA-BstEII digest.
7. Rinse electrophoresed gels in water and photograph with Polaroid 665 film on a UV (300 nm) transilluminator; for quantitation, negatives may be scanned by laser densitometry to record the relative distribution of low molecular weight (LMW) and high molecular weight (HMW) DNA. LMW fragments migrate faster than the 8.45 kb fragment of λDNA-BstEII digest.

C. Results

The endonucleolytic assay permits qualitative evaluation of the configuration of DNA fragmentation (ladder vs. smear) as well as quantitation (percent LMW). Negative controls may include DNA from nonapoptotic sources, and positive controls may include DNA digested with micrococcal nuclease. Nucleosomal laddering has been observed in teratogen-induced cell death.[17,26,27] Typical results are shown for day 8 mouse embryos where large-scale apoptosis was induced by 2′-deoxyadenosine intoxication (Plate 1B). Integration of the area under LMW and HMW densitometric peaks can yield quantitative information on the apoptotic fraction of the total cell population. Since only a small percentage of the total cell population may be apoptotic at any one time, sensitivity is an issue. If necessary, this can be improved by [^{32}P]-dideoxyATP labeling of free 3′-hydroxyl end of the DNA fragments with terminal transferase.[28] Although internucleosomal DNA cleavage is the most characteristic feature of apoptosis, it should not be used as the sole criterion.

V. TUNEL

A. Principle

Detection of large (50 to 300 kbp) and small (185 bp) DNA fragments in histological sections is possible because of the *in situ* terminal transferase reaction.[29,30] This method is based on the specific binding of terminal deoxynucleotidyl transferase (TdT) to free 3′-hydroxyl ends of DNA in tissue sections. TdT catalyzes template-independent synthesis of a deoxynucleotide heteropolymer to protruding, recessed, or blunt 3′-ends of DNA. By incorporating biotin- or digoxygenin-labeled dUTP, the heteropolymer may be localized histochemically. The acronym TUNEL is derived from **t**erminal transferase d**U**TP **n**ick **e**nd **l**abeling.[29] A commercial ApopTag kit is available (Oncor; Gaithersburg, MD).

B. Protocol

1. Embryos are fixed in phosphate-buffered formalin at 4°C overnight and embedded in paraffin by routine methods. Paraffin sections are collected on positively charged microscope slides (Superfrost/Plus, Fisher Scientific).
2. Sections are deparaffinated in 2 × 5 minute xylene, 2 × 5 minute absolute ethanol, 1 × 3 minute 95% ethanol, 1 × 3 minute 70% ethanol, 1 × 3 minute water, and 1 × 5 minute phosphate-buffered saline (PBS).
3. Sections are digested for 15 minutes at room temperature with proteinase K (5 to 20 µg/ml PBS) and washed with 4 × 2 minutes reagent-grade water.
4. Endogenous peroxidase is blocked (if necessary) with 2% hydrogen peroxide in PBS for 5 minutes at room temperature; sections are rinsed 2 × 5 minutes PBS.
5. Equilibrate briefly in 13 µl TdT/equilibration buffer (Oncor) or, alternatively, with 30 m*M* Tris-HCl, pH 7.2, containing 140 m*M* sodium cacodylate and 1 m*M* cobalt chloride, under a 1 cm^2 plastic coverglass at room temperature.
6. Blot excess TdT/equilibration buffer and add 10 µl TdT/reaction buffer. This solution is prepared immediately before use by the addition of 32 µl TdT enzyme in stabilization buffer (Oncor), –20°C, and 76 µl reaction buffer (Oncor, containing digoxygenin-11-dUTP and dATP in an optimal ratio). Cover with plastic cover slip and incubate the slides at 37°C in a humidified chamber for 60 minutes.
7. Stop the reaction by immersing slides in prewarmed stop/wash buffer (Oncor) or, alternatively, with 30 m*M* sodium citrate-300 m*M* sodium chloride at 37°C for 30 minutes with agitation each 10 minutes.
8. Wash 3 × 5 minutes PBS.
9. Apply anti-digoxygenin-peroxidase conjugate (Oncor) to sections and incubate for 30 minutes at room temperature. Wash 3 × 5 minutes with PBS.
10. Color development with ImmunoPure peroxidase substrate (Pierce Chemical Co.). The solution should be prepared immediately before use and filtered at 0.4-µm. Closely monitor the reaction over a 2- to 6-minute period.

I. Introduction

Since flow cytometers first became commercially available in the 1970s, the number used for clinical and research applications has steadily increased, presently nearing 10,000 worldwide.[1] The ability to rapidly analyze thousands of cells to determine phenotype, to quantify cellular components (e.g., DNA, RNA, protein, calcium, etc.), and to isolate rarely occurring cells for subsequent analysis or culture has greatly advanced research in such fields as biotechnology, hematology, immunology, microbiology, oncology, parasitology, pathology, and reproductive biology.

The goals of this chapter are to provide a basic understanding of how flow cytometers work, to discuss some of the important considerations when doing an experiment, and to illustrate some applications of flow cytometry in the study of normal and abnormal development. In addition, we provide detailed experimental procedures for some commonly used techniques and offer a list of suggested references for those in search of additional information.

A. How a Flow Cytometer Works

In essence, particles (e.g., intact cells, nuclei, chromosomes, bacteria) that have been stained with specific fluorescent probes are suspended in a fluid stream and exposed to one or more wavelengths of light (Figure 9.1). The amount of light each particle scatters and emits as fluorescence is then quantified to provide insight into the particle's biophysical and biochemical characteristics (or "parameters", in flow cytometric parlance). Scattered light

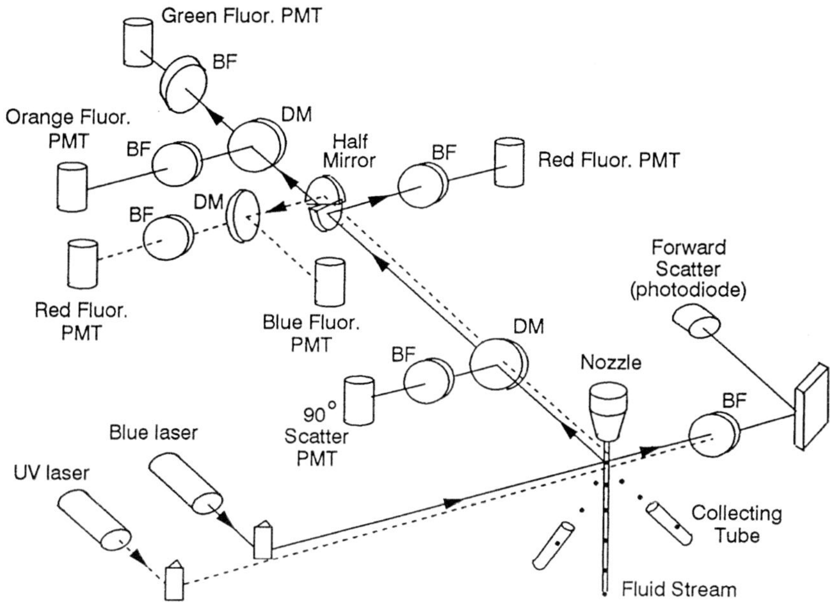

FIGURE 9.1
A schematic of the Becton-Dickinson FACStar Plus illustrates the general design of a research-grade, stream-in-air, dual-laser flow cytometer. Particles stained with fluorescent dyes and suspended in a fluid stream flow past laser beams, and the resulting scattered light and emitted fluorescence are collected by lenses (not shown). Dichroic mirrors (DM) and barrier filters (BF) segregate the scattered/emitted photons by wavelength for detection by photomultiplier tubes (PMTs). In addition, cell subpopulations of interest can be collected via the process of fluorescence-activated cell sorting (FACS).

provides information on such biophysical characteristics as particle size, refractive index, granularity, and density, while the fluorescence emission from stoichiometrically binding fluorochromes or fluorochrome-conjugated antibodies provides a quantitative measure of the content of the cellular components (e.g., DNA, protein, calcium, a specific gene product) to which the probes are bound.

Although all flow cytometers are similar in principle, they range in complexity from easy-to-operate, but inflexible, clinical units to complex research units that can be reconfigured for a wide variety of applications. The simpler clinical units serve well as core instruments to be operated independently by several investigators for such routine analyses as lymphocyte phenotyping or measurement of tumor-cell DNA content. The complex research-grade instruments, in contrast, typically require a dedicated operator to ensure consistent data quality but offer the flexibility to accommodate several lasers for multiple-parameter analysis and to collect cells of interest through the process of fluorescence-activated cell sorting (FACS).

Yet, despite the apparent complexity of the cytometer, the principles of operation are rather simple: the cytometer creates a constant-velocity vertical stream ("sheath") of water or ionic buffer that intercepts one or more light beams that most often originate from a laser but, in some units, from a mercury-arc lamp. The particles of interest are introduced into the center of the fluid stream and, by a process called hydrodynamic focusing, pass single file through the beam(s). The actual point where the particle intercepts the beam may occur within a flat-sided quartz cuvette or after the cylindrical stream has exited a small round orifice (i.e., "stream-in-air" analysis). Upon excitation by the beam(s), the resulting scattered light and emitted fluorescence are collected by a system of mirrors and lenses and directed to a series of photomultiplier tubes (PMTs) for fluorescence and low-intensity scatter quantitation or photodiodes for high-intensity (forward) scatter measurement (Figure 9.1).

Once scattered/emitted photons are detected by the PMTs and photodiodes, the resulting voltage signals are digitized via analog-to-digital convertor (ADC) circuitry and transformed by computer software into one-, two-, or three-parameter graphic displays (Figure 9.2). In one-parameter displays (histograms), the abscissa is typically partitioned into either 256 (2^8) or 1024 (2^{10}) divisions ("channels"), and the relative scatter/fluorescence signals from each of several thousand cells are accumulated into the appropriate channels (Figure 9.2A). For example, in analyzing DNA content, if the cytometer were configured to register a resting diploid cell in channel 300, a mitotic cell containing approximately twice the DNA should fall near channel 600. At each channel, the relative frequency of cells exhibiting that level of scatter/fluorescence is represented by the plot's height.

In two-parameter displays (cytograms), the point representing the magnitude of particle scatter/emission is depicted as a point in a Cartesian plot with one parameter quantified along the abscissa and the second along the ordinate. In an isometric (peak-and-valley) plot (Figure 9.2B.1), the relative frequency of particles exhibiting the same scatter/emission properties is depicted by the height of the three-dimensional peaks. In dot plots (Figure 9.2B.2), frequency is registered as increased pixel intensity and displayed as a color or a grey level, depending on the monitor and printer. In a contour plot (Figure 9.2B.3), frequency is depicted by the number of concentric tracings (isopleths). Three-parameter displays may be represented as a closed cube (Figure 9.2C.1) with each of the three visible surfaces displaying two-parameter dot plots, or alternatively as an open cube (Figure 9.2C.2) with data plotted as clouds of points along the x, y, and z axes.

To further analyze these data, one typically uses software to determine the relative percentages and average signal intensities of clusters or regions of interest (Figure 9.3A.1). Often this is used in conjunction with the powerful technique of gating (Figure 9.3A.2), by which particular cell clusters on one cytogram can be analyzed to determine other fluorescence/scattering characteristics. This is achieved either by providing conditional commands to the computer to display only data meeting certain criteria or by using software

A. One-Parameter Histograms

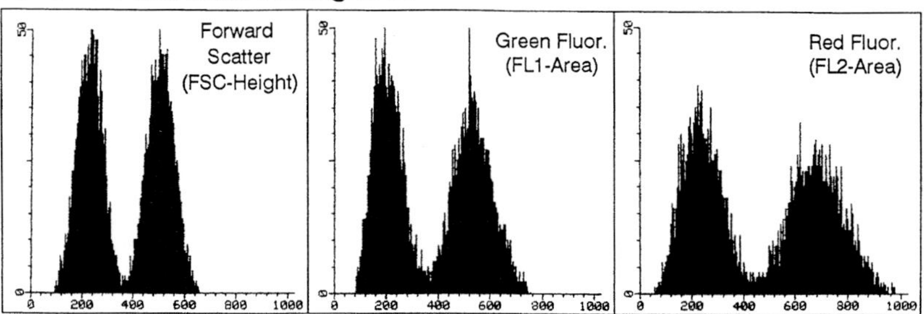

B. Two-Parameter Cytograms

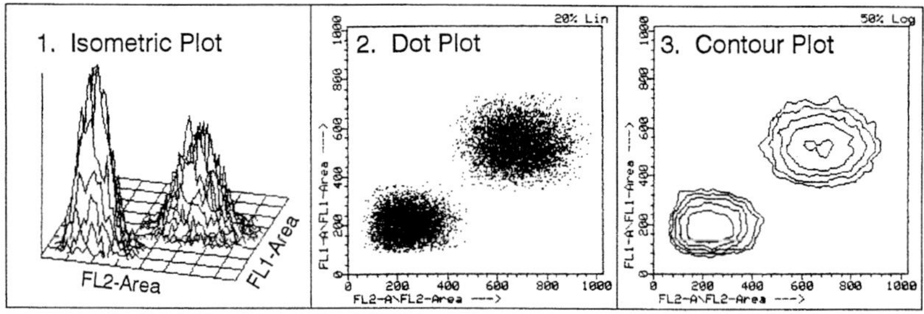

C. Three-Parameter Plots

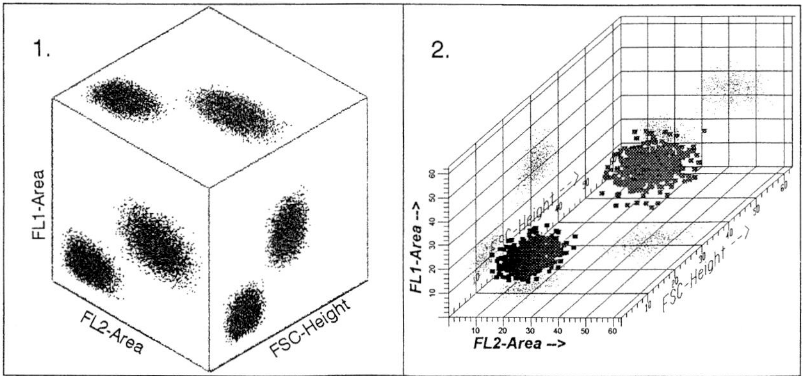

FIGURE 9.2
Flow cytometric data plotted as one-, two-, and three-parameter histograms. In one-parameter histograms (**A**), relative signal intensity is indicated along the abscissa by channel number and relative frequency along the ordinate by plot height. In two-parameter histograms (cytograms) (**B**), the location of clusters along both axes reflects the signal magnitude for each parameter, while frequency is represented as the height of three-dimensional peaks (isometric plot), pixel intensity (dot plot), or the number of concentric tracings (contour plot). In three-parameter plots (**C**), data may be displayed either as multiple two-parameter plots or as clouds within a three-dimensional plot.

A. Two-Parameter Data Analysis

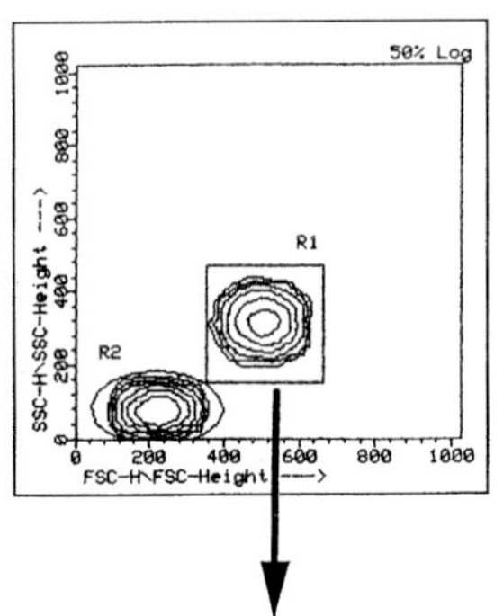

1. Regional Analysis

Parameters: FSC-H(LIN),SSC-H(LIN)
Total= 10000 Gated= 10000

Rgn	Events	% Gated	% Total	Xmean	Ymean
1 R1	5004	50.04	50.04	127.59	80.63
2 R2	4994	49.94	49.94	57.60	21.36

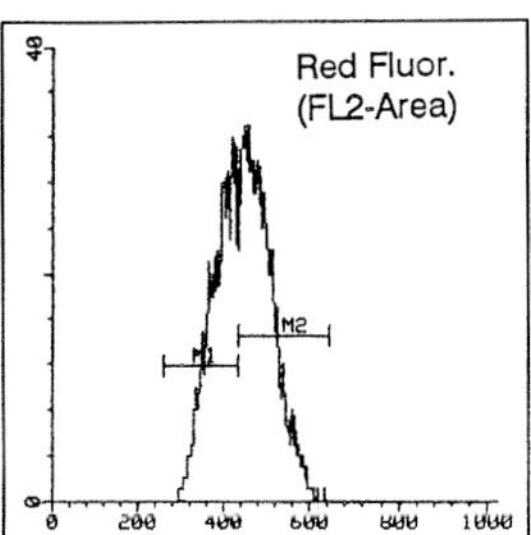

2. Regional Analysis After Gating by Scatter

Selected Preferences: Arithmetic/Linear
Parameter FL2-A FL2-Area Gate G1= R1

M	Left,Right	Events	%	Peak	PkChl	Mean	Median	SD	CV %
0	0, 1023	5012	100.00	33	451.00	444.66	445.00	60.42	13.59
1	263, 437	2283	45.55	32	421.00	390.72	396.00	31.73	8.12
2	437, 645	2757	55.01	33	451.00	489.26	483.00	37.49	7.66

B. Cell-Cycle Analysis

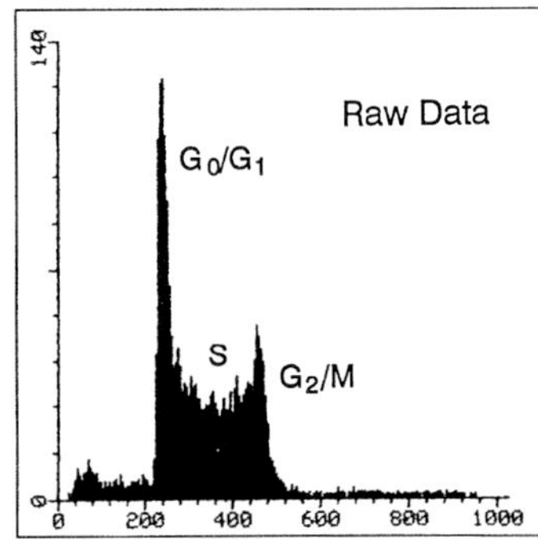

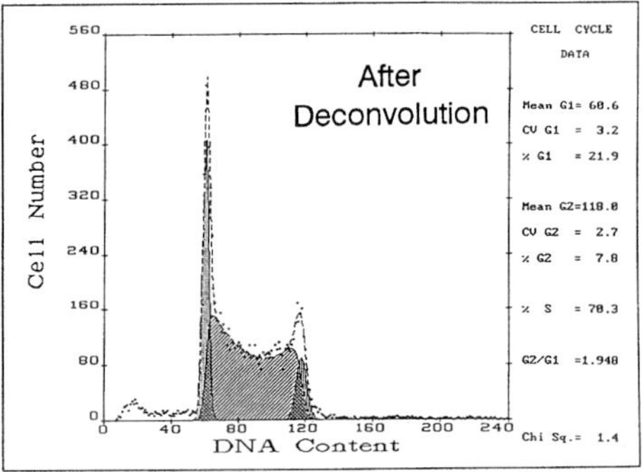

FIGURE 9.3
Flow cytometric data can be quantified by regional analysis to determine the frequency and average signal intensity of cells within clusters of interest (**A1**). By using the technique of gating, one can quickly determine all the parameters of interest for a particular cluster (**A2**). In the case of DNA analysis, one can use software programs to deconvolute the DNA-content histogram into its component cell-cycle phases (**B**).

that color codes each population of interest on all graphic displays. In addition, software also exists for specialized applications, such as deconvoluting DNA-histogram data into the three component cell-cycle phases that are distinguishable by DNA content: G_0/G_1, S, and G_2/M (Figure 9.3B).

B. Factors Influencing Cytometer Performance

1. System Alignment and Configuration

Though straightforward in concept, in actual practice one must take several precautions to guarantee the quality of flow cytometric data. The most important consideration is ensuring proper alignment of all optical elements. By monitoring the fluorescence and scatter signals of standard particles such as chicken erythrocytes that have been made autofluorescent by glutaraldehyde fixation or commercially available beads of uniform size and fluorescence intensity, one can adjust the laser beams, lenses, mirrors, and apertures to ensure optimal focus. Then, by charting the fluorescence and scatter values of the standard particles over time, system performance can be monitored to ensure consistent data quality during and between experiments. In addition, there are standards composed of several populations of beads with different, known intensities (measured as units of molecular equivalent soluble fluorescence, MESF), which can be used to ensure linearity of fluorescence detection and to determine the threshold below which fluorescence cannot be distinguished from optical and electrical noise.

Another important consideration is the effect of sample air pressure, which controls particle flow rate. Typically, one can achieve excellent resolution running samples containing 10^5 to 10^6 cells or nuclei per milliliter at flow rates of 500 to 1000 particles per second. However, if the samples are substantively more dilute, increasing sample pressure to maintain high flow rates will widen the sample stream and allow particles to move off-center. Since the intensity of a laser beam follows a gaussian distribution,[2] these acentrically suspended particles will be exposed to lower intensity light, and their scatter and fluorescence signals will be artifactually decreased. As such, it is important to keep sample air pressure within the manufacturer's recommended ranges and to optimize flow rate by maintaining proper particle concentration when possible.

A less obvious, but no less important, consideration is ensuring that the fluorescence emission from one probe is detected only by the intended PMT(s). To avoid cross-detection by other PMTs, optical filters are used to select desired wavelength bands, typically in the green (510 to 540 nm), yellow (560 to 580 nm), orange (605 to 620 nm), and red (≥620 nm) ranges. For this purpose, there are two basic types of filter elements: dichroic mirrors, which are designed to pass certain wavelengths while reflecting others, and barrier filters, which are designed to pass either a specific band of wavelengths or all wavelengths above (long-pass filter) or below (short-pass filter) a specific cutoff. Thus, by understanding the excitation and emission characteristics of the fluorochromes, one can select the proper combination of optical elements to ensure that only the desired wavelengths are detected. Unfortunately, optical elements are rarely perfect and, more often than not, also pass some light of undesired wavelengths. Thus, compensation circuitry is frequently employed to correct for the spectral overlap arising from multiple dyes and/or from the multiple-wavelength emissions of a single dye.

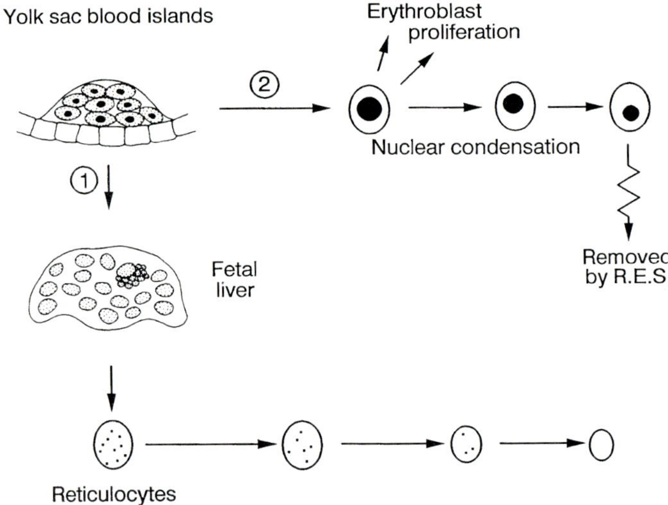

FIGURE 9.11
Fluorescence quantitation can be used to study the effects of toxicants on fetal hematopoiesis. In the rat, hematopoiesis begins in yolk sac islands, which by GD 10 release precursor cells that seed the liver (1) as well as erythroblasts that proliferate in the circulation (2). By GD 14, the liver becomes the major hematopoietic organ, producing reticulocytes, while the circulating erythroblasts cease proliferating, become pyknotic, and are eventually removed by the reticuloendothelial system. Thus, the relative concentration of liver-derived reticulocytes to yolk-sac-derived erythroblasts can be used as an indicator of hematopoietic development.

Höechst dye is absorbed by PI and released as lower energy (higher wavelength) light.[40] The details of this phenomenon are explained elsewhere,[34] but it serves to illustrate that when using more than one dye, one must consider not only the correct optical filter configuration, but how the dyes may interact.

4. Detection of Perturbed Hematopoiesis

In the developing rat, hematopoiesis begins in yolk sac islands (Figure 9.11), which by GD 10 release precursor cells that seed the liver and large nucleated erythroblasts that proliferate in the circulation.[41] With development, the rate of erythroblast proliferation decreases (Figure 9.7), and by GD 14 the erythroblasts become pyknotic and are then removed by the reticuloendothelial system near term. Concurrent with the decreased erythroblast proliferation, the seeded liver becomes the major hematopoietic organ, producing smaller nonnucleated reticulocytes that by term represent essentially the only circulating red blood cells.[42]

Since the nonnucleated reticulocytes are small and contain significant levels of RNA, they can be distinguished not only from the larger, nucleated erythroblasts but also from the smaller nucleic-acid-depleted adult erythrocytes. To measure the relative size differences of these cell types, one can use light scatter (larger particles scatter more light in the forward direction), especially if specimens are first fixed to minimize differences in membrane refractive index. For detecting differences in nucleic acid content, nucleated and nonnucleated cells can be readily distinguished by any of the aforementioned DNA-binding fluorochromes; however, to discriminate the reticulocytes from erythrocytes, the dye must also bind all RNA. Unlike the DNA-specific dyes Höechst 33342 and DAPI, or PI (which binds both DNA and double-stranded RNA), thiazole orange stoichiometrically binds both double and single-stranded RNA as well as DNA,[43] yet it is less susceptible to variations in staining conditions than acridine orange.[44]

Thus, quantifying thiazole-orange fluorescence together with forward scatter (size) permits discrimination of erythroblasts, reticulocytes, and adult erythrocytes (Figure 9.12). This provides a means for monitoring the effects of toxicants on the relative rate of reticulocyte release (Figure 9.12), yet also eliminates the need to sample by fetal cardiac puncture, the standard, but technically difficult, method for obtaining fetal blood samples free of maternal contamination. Moreover, since reticulocyte RNA content normally decreases with maturation, thiazole orange fluorescence on a linear scale (Figure 9.13) can be used to detect deviations in reticulocyte maturation.[46] Using this approach, we demonstrated that 48 hours after maternal injection of 5-FU (a chemotherapeutic agent known to induce anemia)[47] at 40 mg/kg maternal body weight, the fetal liver released, at a reduced rate (Figure 9.12), RNA-depleted reticulocytes (Figure 9.13), followed 24 to 48 hours thereafter by reticulocytes containing elevated levels of RNA.[45] Moreover, by fixing fetal blood in a solution of sodium dodecyl sulfate (SDS) in glutaraldehyde to "sphere" the reticulocytes, then staining with both thiazole orange (RNA/DNA) and Höechst 33342 (DNA), it is possible to discriminate and quantify micronuclei-containing reticulocytes as an indicator of DNA damage during erythropoiesis.[48]

Thus, thiazole orange permits rapid qualitative and quantitative assessment of the progression of fetal hematopoiesis. In this context, it is important to note that the ability to discriminate and quantify the low- and high-RNA reticulocyte subpopulations allowed detection of a phenomenon that methods reporting only average RNA content likely would have missed, for a mean fluorescence value of these two extreme subpopulations would fall within the control range.

B. Immunofluorescence Quantitation

1. Detection of Apoptotic Cells by Labeling DNA Strand Breaks

The ability to label antibodies with a wide range of fluorochromes provides a powerful tool for the detection of one or more surface, nuclear, or cytoplasmic antigens. One application of this technique that has gained wide acceptance is the detection of apoptotic cells by labeling the DNA strand breaks that result from endonuclease-mediated digestion.[49] This relatively simple technique

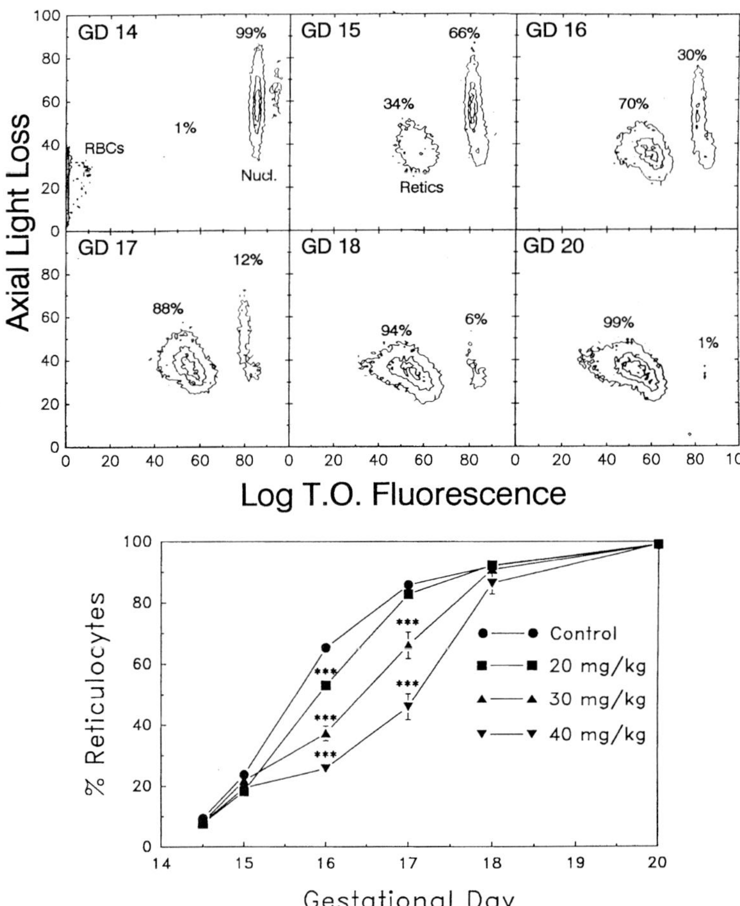

FIGURE 9.12
Using thiazole orange (T.O.) to stain the RNA and DNA of fetal blood cells together with axial light loss (size) allows discrimination of three blood cell populations: nucleated erythroblasts derived from the yolk sac (Nucl.), non-nucleated reticulocytes released from the liver (Retics), and contaminating maternal erythrocytes (RBCs), which were excluded from subsequent analyses. (Logarithmic scaling was required to display nonfluorescent RBCs, low-fluorescence reticulocytes, and highly fluorescent erythroblasts simultaneously on scale.) The ability to monitor the relative percentages of circulating fetal blood cells over time allows detection of toxicant-induced perturbations of erythropoiesis, as illustrated by the dose-dependent reduction in reticulocyte percentage following *in utero* exposure to 5-FU on GD 14 (line graph); *** = $p < 0.001$ compared to control. (From Zucker, R.M. et al., *Teratology,* 51, 37, 1995. With permission.)

employs exogenous terminal deoxynucleotidyl transferase (TdT) to label the 3′-OH ends of the broken strands with digoxigenin-conjugated deoxyuridine triphosphate (dUTP), which is subsequently labeled with an anti-digoxigenin antibody conjugated to fluorescein, thereby staining apoptotic cells green (Figure 9.14). By using PI as a counterstain, one can also identify the DNA content of the apoptotic cells to get some indication of cell-cycle phase specificity.

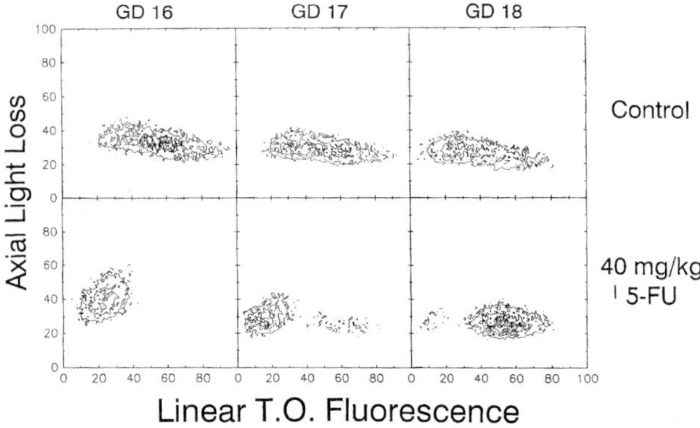

FIGURE 9.13
Linear quantitation of thiazole orange (T.O.) fluorescence following GD-14 maternal exposure to 5-FU reveals that circulating reticulocytes on GD 16 exhibit reduced RNA content, suggestive of inhibited RNA synthesis. Yet 24 to 48 hours thereafter, reticulocytes exhibit elevated RNA content, suggestive of compensatory premature release. (From Zucker, R.M. et al., *Teratology,* 51, 37, 1995. With permission.)

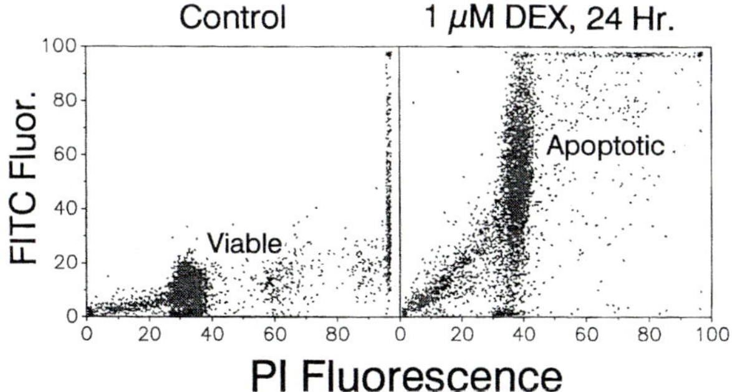

FIGURE 9.14
Immunofluorescence assays can be used to identify apoptotic cells by their characteristic DNA strand breaks. By using exogenous terminal deoxynucleotidyl transferase (TdT) to label breaks with dUTP-digoxigenin and then FITC-conjugated antidigoxigenin antibody, apoptotic cells can be identified by their increased FITC fluorescence.

2. Detection of DNA-Synthesizing Cells by BrdU Incorporation

Another common application of immunofluorescence is the use of bromodeoxyuridine (BrdU) labeling to identify cells actively synthesizing DNA.[50] In this technique, cells are pulse-labeled by exposure to BrdU, which is incorporated in place of thymidine into the DNA of proliferating cells. The cells are isolated, washed, and fixed with 70% ethanol, and the DNA is denatured by treatment with hydrochloric acid to expose the incorporated BrdU, which is then indirectly labeled with a primary mouse-derived antibody

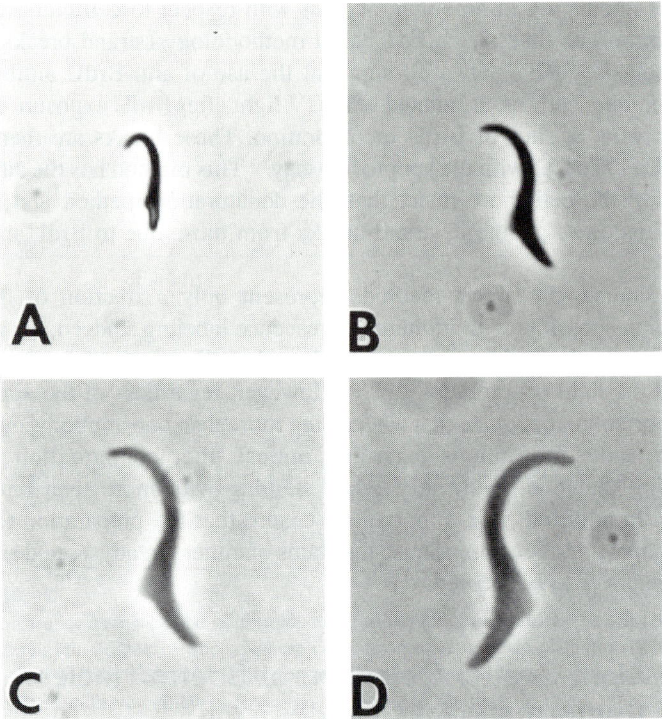

FIGURE 9.16
Rat sperm nuclei treated with sodium dodecyl sulfate and dithiothreitol exhibit a time-dependent increase in size and translucence. Photos taken before (**A**) and 15 (**B**), 30 (**C**), and 40 (**D**) minutes after treatment. (From Zucker, R.M. et al., *Cytometry,* 13, 39, 1992. With permission.)

Upon decondensation, signal intensity increases with the increasing size of the opaque particle (Figures 9.16 and 9.17), but then decreases as the refractive index of the particle decreases and an increasing proportion of incident light is passed rather than scattered.[62] The degree to which this biphasic shift occurs is dependent on both the wavelength of the incident light and the method of detection. Our experience suggests that, as an indicator of the changes in size and refractive index taking place during decondensation, the forward scatter of red (633 nm) light is more sensitive than that of blue (488 nm) light,[61] and that forward scatter detection is more sensitive than axial light loss (extinction). Similar analyses with the morphologically different sperm of several species,[61] however, suggests that axial light loss was more sensitive than forward scatter to differences in sperm nuclear shape.

It also should be noted that, although less sensitive for detecting morphological changes in sperm nuclei, wide-angle (90°) scatter (i.e., light scattered orthogonal to the incident beam; see Figure 9.1) has proved to be very useful in differentiating white blood cells,[63] detecting mitotic nuclei,[12] and detecting altered refractive index as a consequence of toxicant-induced membrane denaturation.[64]

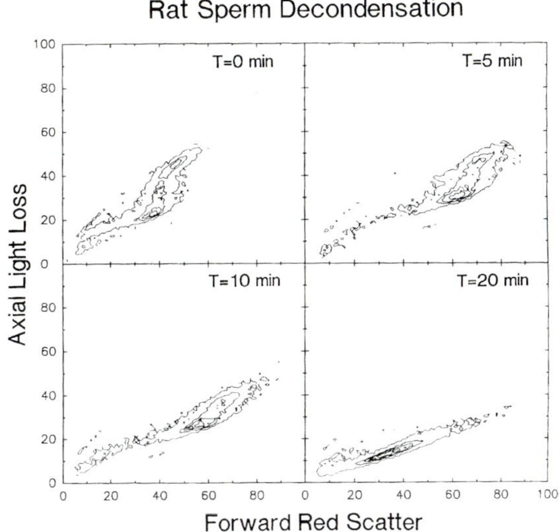

FIGURE 9.17

Using two modes of light-scatter detection — axial light loss (extinction) and forward (small-angle) scatter — changes in rat-sperm nuclear morphology are detected as a biphasic shift in signal intensity. Initially, scatter signals increase with increasing nuclear size but then decrease as the decondensing nuclei become more translucent. Note that the S-shaped distribution at T = 0 results from the relationship of nuclear radial orientation in the sample stream to the amount of light scattered. (From Zucker, R.M. et al., *Cytometry*, 13, 39, 1992. With permission.)

D. Fluorescence-Activated Cell Sorting

Cytometers equipped with sorting electronics offer the capacity to identify and collect purified cell subpopulations. The underlying principal is rather simple: a piezoelectric transducer rapidly vibrates the flow nozzle and breaks the exiting sample stream into uniform microscopic droplets. When a cell meeting the predetermined scatter and/or fluorescence criteria is identified, the cytometer places a small charge on the droplet containing the desired cell. As all the droplets pass by two charged deflection plates, the desired (and now charged) droplet is deflected away from the other droplets composing the waste stream and can thereby be collected for subsequent processing.

The sorting capability is routinely used to correlate a cell population's light scattering and fluorescence characteristics with its microscopic appearance. In addition, this option offers the ability to conduct further investigations with homogeneous cell populations. For example, by dissociating embryonic cells and using an FITC-conjugated monoclonal antibody, Stemple and Anderson[65] were able to identify and isolate neural crest cells for subsequent culture.

Although simple in principal, it should be noted that droplet sorting requires substantial expertise to identify and manage the numerous factors

B. Conclusion

As can be seen by this brief overview, flow cytometry offers a variety of capabilities for investigating the cellular and subcellular mechanisms that underlie both normal and abnormal development. The advantages include the capacity for rapid, multiparameter analysis and thus the ability not only to obtain data from each of several thousand cells, but to determine the relationships between the various characteristics exhibited by each cell. However, as with any technology, there are also limitations. The most notable drawbacks include: (1) the requirement for single-cell suspensions and thus the loss of tissue architecture, (2) the interference of autofluorescence in the quantification of low-level fluorescent probes, (3) the difficulty in identifying clusters of interest in units lacking sorting capability, (4) the considerable up-front expense for the equipment ($75,000 for an analyzer, $200,000 for a sorter), and (5) the cost and time needed for training, particularly for sophisticated research instruments requiring a dedicated operator. Yet, once these limitations are understood, the technology offers the investigator unique capabilities that are limited only by one's imagination.

IV. Selected Experimental Procedures

A. Cell Cycle Analysis of Embryonic Tissues[20]

1. Reagents

   Lysing cocktail: 0.2% Nonidet P-40 (Sigma #N6507; St. Louis, MO) in phosphate-buffered saline (PBS, Life Technologies #310-4080; Grand Island, NY), pH 7.4, supplemented with 0.5 mg/ml RNase A (Sigma #R4875).

   Stain: 1 mg/ml propidium iodide (PI, Sigma #P4170) in PBS. Keep cold and protect from light.

2. Procedure

   a. Mince tissues of interest into 1 to 2 mm cubes and dissociate cells by expressing 2 to 3 times through a 26-g needle.

   b. A sample number of 0.2 to 2×10^6 cells is optimal. For dense suspensions, remove a 20 to 100-μl aliquot to a fresh tube. For dilute suspensions, transfer cells to a 12 × 75-mm culture tube, centrifuge at 200× g for 10 minutes at room temperature, and decant the supernatant.

   c. Add cocktail to a total volume of 0.5 ml and vortex. Incubate at room temperature for 30 minutes, then chill on ice.

   d. Add 25 μl stain, vortex, incubate for 15 minutes on ice, then analyze at 0°C. Collect red fluorescence of 10,000 nuclei using 488 nm excitation, a 630-nm Schott barrier filter, and doublet discrimination.

B. Detection of Thymocytic Apoptosis

Preparation of Thymocyte Cultures

1. Reagents (prewarm all reagents to 37°C)

   Culture medium: RPMI 1640 medium (Life Technologies #11875-044; Grand Island, NY) with 10% fetal bovine serum (FBS, Life Technologies #16140-014), 25 m*M* HEPES (Sigma #H3375; St. Louis, MO), and 50 μg/ml gentamicin (Life Technologies #600-5750). Make fresh.

   PBS: Filter (0.45 μm) phosphate-buffered saline (PBS; Life Technologies #310-4080), pH 7.4.

   FBS/PBS: Add 1 ml FBS to 9 ml PBS and sterilize by filtration.

2. Procedure

   a. Remove thymus from a 3- to 6-week-old male rat, wash off blood with PBS, remove mediastinum, and transfer to petri dish containing 2 to 3 ml FBS/PBS.

   b. Mince thymus into 1 to 2 mm cubes, transfer to 15-ml tube, and triturate repeatedly with a transfer pipet. Filter through gauze.

   c. Determine cell density with a Coulter Counter or hemacytometer.

   d. Dilute thymocyte suspension in culture medium to a concentration of ~2 × 10^6 cells per milliliter and incubate 1 to 6 hours with an apoptosis-inducing agent (e.g., 1 μ*M* DEX or TBT) in a humidified atmosphere of 5% CO_2.

Nuclear Apoptosis Assay[33]

1. Reagents

   Lysing buffer: 0.2% Nonidet P-40 (Sigma #N6507; St. Louis, MO) in 0.85% NaCl (Sigma #S9625) with 25 m*M* HEPES (Sigma #H3375), pH 7.4.

   Stain: 1 mg/ml propidium iodide (PI, Sigma #P4170) in PBS. Keep cold and protect from light.

2. Procedure

   a. Transfer 1 ml of 2 × 10^6 cells/ml thymocyte suspension to 12 × 75-mm culture tube and centrifuge at 200× g for 10 minutes at room temperature.

   b. Decant supernatant. Add 0.5 ml lysing buffer, vortex, and incubate at room temperature for 15 minutes.

   c. Chill on ice, add 25 μl stain, vortex, and incubate on ice for 15 minutes. Analyze red fluorescence at 0°C with 488 nm excitation and a 630-nm Schott barrier filter.

Cellular Apoptosis Assay[34]

1. Reagents

   Stains: 1.13 mg/ml Höechst 33342 (Sigma #B2261; St. Louis, MO) or 1.06 mg/ml Höechst 33258 (Sigma #B2883) in deionized H_2O. 1 mg/ml propidium iodide (PI, Sigma #P4170) in PBS. Keep all stains cold and protected from light.

2. Procedure
   a. Transfer 1 ml of 2×10^6 cells per milliliter thymocyte suspension to a $12 \times$ 75-mm culture tube. Add 5 µl of either Höechst stain.
   b. Vortex. Incubate for 10 minutes in a 37°C water bath.
   c. Chill on ice, add 5 µl PI, vortex, incubate for 15 minutes, and analyze at 0°C.
   d. Höechst stains are UV excitable (351 to 363 nm) with peak emission in the blue region (475 to 485 nm). PI is detected with a 630-nm Schott barrier filter. Read Höechst 33258 fluorescence on a logarithmic scale; Höechst 33342 fluorescence on a linear scale.

C. Detection of Perturbed Hematopoiesis

Blood Sample Acquisition

1. Reagents

   PBS: Filter (0.45 µm) phosphate-buffered saline (PBS; Life Technologies #310-4080; Grand Island, NY), pH 7.4, prewarmed to 37°C.

   Heparinized FBS/PBS: Add 1 ml FBS and 1750 units heparin (Sigma #H3125; St. Louis, MO) to 9 ml PBS, sterilize by filtration, and warm to 37°C.

   Fixative: 5% paraformaldehyde (Sigma #P6148; St. Louis, MO) in PBS.

2. Procedure
   a. Remove conceptus (including placenta), wash in plain PBS, and place in a small petri dish containing 1 ml heparinized PBS/FBS.
   b. Rupture vitelline and umbilical vessels and let fetus exsanguinate while mixing blood with the surrounding medium to prevent clotting.
   c. Transfer diluted blood to a 12×75-mm culture tube and mix with an equal volume of fixative.
   d. Store at 5°C until ready for analysis.

Flow Cytometric Analysis[45]

1. Reagents

   Stain: 0.1 µg/ml thiazole orange (Retic-COUNT, Becton-Dickinson #92-0004; San Jose, CA).

2. Procedure
   a. Determine sample density with Coulter Counter or hemacytometer.
   b. Transfer 1×10^6 cells to a 12×75-mm culture tube and add 1 ml thiazole orange.
   c. Incubate in the dark for 30 minutes at room temperature.

d. Analyze at 0°C using 488-nm excitation and the standard FITC filter set for emission. Use log scale to determine relative percentage of reticulocytes and erythroblasts; linear scale to quantitate reticulocyte RNA content.

D. BrdU Incorporation[25]

1. Reagents

   BrdU: Add 3.1 mg/ml 5-bromo-2′-deoxyuridine (BrdU, Sigma #B5002; St. Louis, MO) to 10 ml phosphate-buffered saline (PBS, Life Technologies #310-4080; Grand Island, NY), pH 7.4. Filter, sterilize, and store aliquots at –20°C. Protect from fluorescent light.

   4 *N* HCl/Tween: Add 0.5 ml Tween 20 (Sigma #P1379) to 100 ml 4 *N* HCl. Use fresh.

   PBS/Tween: Add 0.5 ml Tween 20 to 100 ml PBS.

   PBS/Tween/BSA: Add 0.5 g fraction V bovine serum albumin (BSA, #2293-01, Armour Pharmaceutical Co.; Kankakee, IL) and 0.5 ml Tween 20 to 100 ml PBS.

   Antibodies

   Anti-BrdU (Becton-Dickinson #7580; San Jose, CA).

   FITC-conjugated goat-anti-mouse IgG $F(ab')_2$ (Boehringer-Mannheim, #605-29; Indianapolis, IN). Store at –20°C in 20-μl aliquots; protect from light. Dilute 1:100 in PBS/Tween/BSA before use.

   Stain: 1 mg/ml propidium iodide (PI, Sigma #P4170) in PBS. Keep cold and protect from light.

2. Procedure:

   a. Perform assay in incandescent light only. Expose cells to 100 μl BrdU per 10 ml medium for 30 minutes before harvesting.

   b. Centrifuge cells 500× g for 5 minutes. Decant. Wash twice with PBS.

   c. Thoroughly vortex pellet, then slowly add 1 ml of 70% ice-cold ethanol with continuous vortexing. Seal and refrigerate overnight.

   d. Centrifuge cells as above, decant ethanol. Vortex pellet, and add 1 ml HCl/Tween. Incubate at room temperature for 30 minutes.

   e. Wash 3 times with 2 to 3 ml PBS/Tween. Add 100 μl PBS/Tween/BSA to pellet and resuspend thoroughly.

   f. Add 20 μl anti-BrdU, vortex, and incubate 30 minutes at room temperature.

   g. Wash 3 times with 2 to 3 ml PBS/Tween/BSA. Add 100 μl of dilute FITC-antibody to pellet, incubate 30 minutes at room temperature.

   h. Wash 3 times with 2 to 3 ml PBS/Tween/BSA. To final pellet, add 1 ml PBS/Tween/BSA and chill on ice.

   i. Add 50 μl stain, vortex, incubate 5 minutes on ice, and analyze at 0°C. Use 488-nm excitation with standard FITC emission filter and 630-nm Schott barrier filter for PI. Use single-stained samples to set compensation circuitry for emission crossover.

E. Detection of Sperm Nuclear Decondensation by Light Scatter[61]

1. Reagents

   Tris: 50 m*M* Tris buffer (Sigma #T6664; St. Louis, MO), pH 8.0.

   SDS: Stock solution of 3% sodium dodecyl sulfate (SDS, Sigma #L4509; St. Louis, MO) in 50 m*M* Tris buffer.

   DTT: Stock solution of 50 m*M* dithiothreitol (DTT, Sigma #D0632) in Tris.

2. Procedure
   a. Collect sperm nuclei from cauda epididymis of adult rats, sonicate ~10 seconds in Tris to dislodge tails. Centrifuge suspensions through a two-step sucrose gradient to separate tail-free nuclei.
   b. Wash nuclei 3 times in Tris (2500× g for 10 minutes). Transfer 1×10^7 nuclei to a 12 × 75-mm culture tube and dilute to 1 ml with Tris.
   c. To decondense, add 0.25 ml 3% SDS and 0.14 ml 50 m*M* DTT, vortex, and analyze at ambient temperature over time.
   d. Reported data were collected with an Ortho 50H flow cytometer that was configured to quantify the axial light loss (extinction) and forward scatter from a helium-neon (red) laser.

Acknowledgment

The authors thank Dr. Michael Narotsky for his helpful comments and critical review of the manuscript. This document has been reviewed by the National Health Effects and Environmental Research Laboratory, U.S. Environmental Protection Agency, and approved for publication. Approval does not signify that the contents necessarily reflect the views and policies of the Agency, nor does mention of trade names or commercial products constitute endorsement or recommendation for use.

Suggested Readings

The interested reader unfamiliar with the operation and application of flow cytometers may wish to review the following texts.

1. Givan, Alice Longobardi, *Flow Cytometry: First Principles,* Wiley-Liss, New York, 1992. A thorough and highly comprehensible introductory text covering the basics of flow cytometry. In a conversational writing style, Dr. Givan discusses such relevant areas as optics, fluidics, electronics, and data analysis, as well as the principles behind basic clinical and research techniques in a clear, concise, and often entertaining manner.
2. Watson, James V., *Introduction to Flow Cytometry,* Cambridge University Press, New York, 1991. A more detailed treatment of the principles involved in flow cytometry, including the physics of optics, electronics, and hydrodynamics.

Clearly written, although somewhat advanced in scope for the beginner, it will provide valuable insight to those who have already gotten their feet wet (especially if as a result of faulty fluidics!).

3. Shapiro, H., *Practical Flow Cytometry,* 3rd ed., Wiley-Liss, New York, 1995. Considered by many to be the "bible of flow cytometry", this extensive, though at times highly technical, text is useful as a complete reference for questions regarding all of the inner workings of the cytometer, and is also an excellent source for information on frequently used fluorochromes and potential avenues of research.
4. Melamed, M.R., Lindmo, T., and Mendelsohn, M.L., Eds., *Flow Cytometry and Sorting,* 2nd ed., Wiley-Liss, New York, 1990. This 800-page compilation of chapters written by eminent authorities in the fields of cell biology, immunology, and oncology provides an in-depth overview of how flow cytometers work, as well as numerous general and specialized applications. Often highly technical, it best serves as a resource for the many issues one will encounter while obtaining and interpreting flow cytometric data.
5. Robinson, J.P. (Ed.), Darzynkiewicz, Z., Dean, P., Dressler, L., Tanke, H., and Wheeless, L (Assoc. Eds.), *Handbook of Flow Cytometry Methods,* Wiley-Liss, New York, 1993. A handy, well organized collection of guidelines for the preparation of samples for flow cytometric analysis and "cookbook" procedures submitted by several laboratories covering such areas as DNA analysis, phenotyping, and various cellular function assays. The text also contains useful charts on the excitation and emission characteristics of commonly used dyes, as well as a list of suppliers and typical prices.
6. Darzynkiewicz, Z., Robinson, J.P., and Crissman, H.A., Eds., *Methods in Cell Biology: Flow Cytometry,* 2nd ed., Vols. 41 and 42, Academic Press, New York, 1994. At twice the size of the original 1990 edition, this latest compilation covers in 71 chapters the full gamut of applications involving model systems from viruses and bacteria to cells from both the plant and animal kingdoms. In addition to step-by-step procedures, each chapter provides a good overview of the theoretical bases for the described techniques. An exceptional resource that should appeal to both cell and molecular biologists.
7. Mayall, B.H., Ed.-in-Chief, *Cytometry,* The Journal of the International Society of Analytical Cytology, Wiley-Liss, San Francisco, CA. The trade journal of flow cytometry, offering the latest innovations in clinical and research applications.
8. Haugland, R.P., *Handbook of Fluorescent Probes and Research Chemicals,* Molecular Probes, Inc., Eugene OR, 1994. This, the most comprehensive catalog of its kind, provides extensive information about the applications and physical, binding, and optical characteristics of over 1800 fluorescent probes. Regardless of the vendor you use, this book represents a valuable resource for product information.
9. The *Melles Griot Optics Guide* (Irvine, CA). A comprehensive guide to optical filters, mirrors, lenses, and lasers. Each section offers a concise review of basic optical principles and enough spectral charts to assist in selecting the proper combination of elements to customize your cytometer for any application.

References

1. Shapiro, H., *Practical Flow Cytometry,* 3rd ed., Wiley-Liss & Sons, New York, 1995, 29.
2. Shapiro, H., *Practical Flow Cytometry,* 3rd ed., Wiley-Liss & Sons, New York, 1995, 107.
3. Visscher, D.W. and J.D. Crissman, Dissociation of intact cells from tumors and normal tissues, in *Methods in Cell Biology: Flow Cytometry,* Vol. 41, Part A, Darzynkiewicz, Z., J.P. Robinson, and H.A. Crissman, Eds., Academic Press, San Diego, 1994, 1–15.
4. Tom, P.P.L., Studying development in embryo fragments, in *Postimplantation Mammalian Embryos: A Practical Approach,* Copp, A.J. and D.L. Cockroft, Eds., Oxford University Press, Oxford, 1990, 317–337.
5. Freshney, R.I., *Culture of Animal Cells: A Manual of Basic Techniques,* 3rd ed., Wiley-Liss & Sons, New York, 1994, 133–148.
6. Darzynkiewicz, Z. and J. Kapuscinski, Acridine orange: a versatile probe of nucleic acids and other cell constituents, in *Flow Cytometry and Sorting,* 2nd ed., Melamed, M.R., T. Lindmo, and M.L. Mendelsohn, Eds., Wiley-Liss, New York, 1990, 291–314.
7. Baserga, R., *The Biology of Cell Reproduction,* Harvard University Press, Cambridge, MA, 1985, 3–21.
8. Darnell, J., H. Lodish, and D. Baltimore, *Molecular Cell Biology,* Scientific American Books, New York, 1986, 1038–1040.
9. Krishan A., Rapid flow cytofluorometric analysis of mammalian cell cycle by propidium iodide staining, *J. Cell Biol.,* 66, 188–193, 1975.
10. Crissman, H.A. and G.T. Hirons, Staining of DNA in live and fixed cells, in *Methods in Cell Biology: Flow Cytometry,* Vol. 41, Part A, Darzynkiewicz, Z., J.P. Robinson, and H.A. Crissman, Eds., Academic Press, San Diego, 1994, 196–209.
11. Vindelov, L.L., I.J. Christensen, G. Jensen, and N.I. Nissen, A detergent-trypsin method for the preparation of nuclei for flow cytometric DNA analysis, *Cytometry,* 3, 323–328, 1983.
12. Zucker, R.M., K.H. Elstein, R.E. Easterling, and E.J. Massaro, Flow cytometric discrimination of mitotic nuclei by right-angle light scatter, *Cytometry,* 9, 226–231, 1988.
13. Crissman, H.A. and J.A. Steinkamp, Rapid, simultaneous measurement of DNA, protein, and cell volume in single cells from large mammalian cell populations, *J. Cell Biol.,* 59, 766–771, 1973.
14. Gong, J.M., X. Li, F. Traganos, and Z. Darzynkiewicz, Expression of G_1 and G_2 cyclins measured in individual cells by multiparameter flow cytometry: a new tool in the analysis of the cell cycle, *Cell Prolif.,* 27, 357–371, 1994.
15. Roti Roti, J.L., R. Higashikubo, O.C. Blain, and N. Uygur, Cell cycle position and nuclear protein content, *Cytometry,* 3, 91–96, 1982.
16. Darzynkiewicz, Z., F. Traganos, and M.R. Melamed, New cell cycle compartments identified by multiparameter flow cytometry, *Cytometry,* 1, 98–108, 1980.

17. Darzynkiewicz, Z., J. Kapuscinski, F. Traganos, and H.A. Crissman, Applications of pyronin Y in cytochemistry of nucleic acids, *Cytometry,* 8, 138–145, 1987.
18. Traganos, F., Z. Darzynkiewicz, T. Sharpless, and M.R. Melamed, Simultaneous staining of ribonucleic and deoxyribonucleic acid in unfixed cells using acridine orange in a flow cytometric system, *J. Histochem. Cytochem.,* 25, 46–56, 1977.
19. Mirkes, P.E., Cyclophosphamide teratogenesis: a review, *Teratogen. Carcinogen. Mutagen.,* 5, 75–88, 1985.
20. Chernoff, N., J.M. Rogers, A.J. Alles, R.M. Zucker, K.H. Elstein, E.J. Massaro, and K.K. Sulik, Cell cycle alterations and cell death in cyclophosphamide teratogenesis, *Teratogen. Carcinogen. Mutagen.,* 9, 199–209, 1989.
21. Francis, B.M., J.M. Rogers, K.K. Sulik, A.J. Alles, K.H. Elstein, R.M. Zucker, E.J. Massaro, M.B. Rosen, and N. Chernoff, Cyclophosphamide teratogenesis: evidence for compensatory responses to induced cellular toxicity, *Teratology,* 42, 473–482, 1990.
22. Rogers, J.M., B.M. Francis, K.K. Sulik, A.J. Alles, E.J. Massaro, R.M. Zucker, K.H. Elstein, M.B. Rosen, and N. Chernoff, Cell death and cell cycle perturbation in the developmental toxicity of the demethylating agent 5-aza-2′deoxycytidine, *Teratology,* 50, 332–339, 1994.
23. Shuey, D.L., C. Lau, T.R. Logsdon, R.M. Zucker, K.H. Elstein, M.G. Narotsky, R.W. Setzer, R.J. Kavlock, and J.M. Rogers, Biologically-based dose-response modeling in developmental toxicology: biochemical and cellular sequelae of 5-fluorouracil exposure in the rat fetus, *Toxicol. Appl. Pharmacol.,* 126, 129–144, 1994.
24. Pinendo, H.M. and G.F.J. Peters, Fluorouracil: biochemistry and pharmacology, *J. Clin. Oncol.,* 6, 1653–1664, 1988.
25. Elstein, K.H., R.M. Zucker, D.L. Shuey, C. Lau, N. Chernoff, and J.M. Rogers, Utility of the murine erythroleukemic cell (MELC) in assessing mechanisms of action of DNA-active developmental toxicants: application to 5-fluorouracil, *Teratology,* 48, 75–87, 1993.
26. Elstein, K.H., R.M. Zucker, J.E. Andrews, M. Ebron-McCoy, D.L. Shuey, and J.M. Rogers, Effects of developmental stage and organ type on embryo-fetal DNA distributions and 5-fluorouracil-induced cell-cycle perturbations, *Teratology,* 48, 355–363, 1993.
27. Kerr, J.F.R. and B.V. Harmon, Definition and incidence of apoptosis: an historical perspective, in *Apoptosis: The Molecular Basis of Cell Death,* Tomei, L.D. and F.O. Cope, Eds., Cold Spring Harbor Laboratory Press, Plainview, NY, 1991, 5–25.
28. Sen, S., Programmed cell death: concept, mechanism, and control, *Biol. Rev.,* 67, 287–319, 1992.
29. Kerr, J.F.R., Introduction: definition of apoptosis and overview of its incidence, in *Programmed Cell Death: The Cellular and Molecular Biology of Apoptosis,* Lavin, M. and D. Watters, Eds., Harwood Academic Publishers, Switzerland, 1993, 1–15.

30. Darzynkiewicz, Z., S. Bruno, G. Del Bino, W. Gorczyca, M.A. Hotz, P. Lassota, and F. Traganos, Features of apoptotic cells measured by flow cytometry, *Cytometry,* 13, 795–808, 1992.
31. Schwartzman, R.A. and J.A. Cidlowski. Apoptosis: the biochemistry and molecular biology of programmed cell death, *Endocr. Rev.,* 14, 133–149, 1993.
32. Compton, M.M., J.S. Haskill, and J.A. Cidlowski. Analysis of glucocorticoid actions on rat thymocyte deoxyribonucleic acid by fluorescence-activated flow cytometry, *Endocrinology,* 12, 2158–2164, 1988.
33. Zucker, R.M., K.H. Elstein, D.J. Thomas, and J.M. Rogers, Tributyltin and dexamethasone induce apoptosis in rat thymocytes by two mutually antagonistic mechanisms, *Toxicol. Appl. Pharmacol.,* 127, 163–170, 1994
34. Elstein, K.H. and R.M. Zucker, Comparison of cellular and nuclear flow cytometric techniques for discriminating apoptotic subpopulations, *Exp. Cell Res.,* 211, 322–331, 1994.
35. Freshney, R.I., *Culture of Animal Cells: A Manual of Basic Techniques,* 3rd ed., Wiley-Liss & Sons, New York, 1994, 288–289.
36. Hamori, E., D.J. Arndt-Jovin, B.G. Grimwade, and T.M. Jovin, Selection of viable cells with known DNA content, *Cytometry,* 1, 132–135, 1980.
37. Dive, C., C.D. Gregory, D.J. Phipps, D.L. Evans, A.E. Milner, and A.H. Wyllie, Analysis and discrimination of necrosis and apoptosis (programmed cell death) by multiparameter flow cytometry, *Biochim. Biophys. Acta,* 1133, 275–285, 1992.
38. Krishan, A., Effect of drug efflux blockers on vital staining of cellular DNA with Höechst 33342, *Cytometry,* 8, 642–645, 1987.
39. Ormerod, M.G., X.-M. Sun, R.T. Snowden, R. Davies, H. Feranhead, and G.M. Cohen, Increased membrane permeability of apoptotic thymocytes: a flow cytometric study, *Cytometry,* 14, 595–602, 1993.
40. Crissman, H.A. and J.A. Steinkamp, Cytochemical techniques for multivariate analysis of DNA and other cellular constituents, in *Flow Cytometry and Sorting,* 2nd ed., Melamed, M.R., T. Lindmo, and M.L. Mendelsohn, Eds., Wiley-Liss, New York, 1990, 233.
41. Russell, S.R. and S.E. Bernstein, Blood and blood formation, in *Biology of the Laboratory Mouse,* 2nd ed., Green, E.L., Ed., Dover Publications, New York, 1966, 351–372.
42. Bannerman, R.M., Hematology, in *The Mouse in Biomedical Research,* Vol. III. *Normative Biology, Immunology, and Husbandry,* Foster, H.L., J.D. Small, and J.G. Fox, Eds., Academic Press, San Diego, 1983, 294–312.
43. Corash, L., M. Rheinschmidt, S. Lieu, P. Meers, and E. Brew, Enumeration of reticulocytes using fluorescence-activated flow cytometry, *Pathol. Immunopathol. Res.,* 7, 381–394, 1988.
44. Schmitz, F.J. and E. Werner, Optimization of flow-cytometric discrimination between reticulocytes and erythrocytes, *Cytometry,* 7, 439–444, 1986.
45. Zucker, R.M., K.H. Elstein, D.L. Shuey, and J.M. Rogers, Flow cytometric detection of abnormal fetal erythropoiesis: application to 5-fluorouracil-induced anemia, *Teratology,* 51, 37–44, 1995.

46. MacGregor, J.T., P.R. Heniken, L. Whitehand, and C.M. Wehr, The fetal blood erythrocyte micronucleus assay: classification of RNA-positive erythrocytes into two age populations by RNA aggregation state, *Mutagenesis,* 4, 190–199, 1989.
47. Shuey, D.L., R.M. Zucker, K.H. Elstein, and J.M. Rogers, Fetal anemia following maternal exposure to 5-fluorouracil in the rat, *Teratology,* 49, 311–319, 1994.
48. Grawé, J., G. Zetterberg, and H. Amnéus, Flow-cytometric enumeration of micronucleated polychromatic erythrocytes in mouse peripheral blood, *Cytometry,* 13, 750–758, 1992.
49. Gorczyca, W., J. Gong, and Z. Darzynkiewicz, Detection of DNA strand breaks in individual apoptotic cells by the *in situ* terminal deoxynucleotidyl transferase and nick translation assays, *Cancer Res.,* 53, 1945–1951, 1993.
50. Dolbeare, F., H.G. Gratzner, M.G. Pallavicini, and J.W. Gray. Flow cytometric measurement of total DNA content and incorporated bromodeoxyuridine, *Proc. Natl. Acad Sci., U.S.A.,* 80, 5573–5577, 1983.
51. Carlton, J.C., N.H.A. Terry, and R.A. White, Measuring potential doubling times of murine tumors using flow cytometry, *Cytometry,* 12, 645–650, 1991.
52. Ritter, M.A., J.F. Fowler, Y.J. Kim, K.W. Gilchrist, L.W. Morrissey, and T.J. Kinsella, Tumor cell kinetics using two labels and flow cytometry, *Cytometry,* 16, 49–58, 1994.
53. Li X., F. Traganos, M.R. Melamed, and Z. Darzynkiewicz, Detection of 5-bromo-2-deoxyuridine incorporated into DNA by labeling strand breaks induced by photolysis (SBIP), *Intl. J. Oncol.,* 4, 1157–1161, 1994.
54. Li X., F. Traganos, and Z. Darzynkiewicz, Simultaneous analysis of DNA replication and apoptosis during treatment of HL-60 cells with camptothecin and hyperthermia and mitogen stimulation of human lymphocytes, *Cancer Res.,* 54, 4289–4293, 1994.
55. Schlossman, S.F., Ed. *Leucocyte Typing V: White Cell Differentiation Antigen,* Oxford University Press, Oxford, 1994.
56. Schlossman, S.F., L. Boumsell, W. Gilks, J.M. Harlan, T. Kishimoto, C. Morimoto, J. Ritz, S. Shaw, R.L. Silverstein, T.A. Springer, T.F. Tedder, and R.F. Todd, Update: CD antigens 1993, *J. Immunol.,* 152, 1–2, 1994.
57. Gottlieb, M.S., R. Schroff, H.M. Schanker, J.D. Weisman, P.T. Fan, R.A. Wolf, and A. Saxon, Pneumocystis carinii pneumonia and mucosal candidiasis in previously healthy homosexual men: evidence of a new acquired cellular immunodeficiency, *N. Engl. J. Med.,* 305, 1425–1431, 1981.
58. Salzman, G.C., S.B. Singham, R.G. Johnston, and C.F. Bohren, Light scattering and cytometry, in *Flow Cytometry and Sorting,* 2nd ed., Melamed, M.R., Lindmo, T., and Mendelsohn, M.L., Eds., Wiley-Liss, New York, 1990, 81–107.
59. Shapiro, H., *Practical Flow Cytometry,* 3rd ed., Wiley-Liss, New York, 1995, 80.
60. Wolgemuth, D.J., Synthetic activities of the mammalian early embryo: molecular and genetic alterations following fertilization, in *Mechanism and Control of Animal Fertilization,* Hartmann, J.F., Ed., Academic Press, New York, 1983, 415–452.
61. Zucker, R.M., S.D. Perreault, and K.H. Elstein, Utility of light scatter in the morphological analysis of sperm, *Cytometry,* 13, 39–47, 1992.

62. Perreault, S.D., R.R. Barbee, K.H. Elstein, R.M. Zucker, and C.L. Keefer, Interspecies differences in the stability of mammalian sperm nuclei assessed *in vivo* by sperm microinjection and *in vitro* by flow cytometry, *Biol. Repro.*, 39, 157–167, 1988.
63. Salzman G.C., J.M. Crowell, and J.C. Martin, Cell classification by laser light scattering: identification and separation of unstained leukocytes, *Acta Cytol.*, 19, 374–377, 1975.
64. Zucker, R.M., K.H. Elstein, R.E. Easterling, H.P. Ting-Beall, J.W. Allis, and E.J. Massaro, Effects of tributyltin on biomembranes: alteration of flow cytometric parameters and inhibition of Na^+, K^+-ATPase two-dimensional crystallization, *Toxicol. Appl. Pharmacol.*, 96, 393–403, 1988.
65. Stemple, D.L. and D.J. Anderson, Isolation of a stem cell for neurons and glia from the mammalian neural crest, *Cell,* 71, 973–985, 1992.
66. Shapiro, H., *Practical Flow Cytometry,* 3rd ed., Wiley-Liss, New York, 1995, 223.
67. Watson, J.V., *Introduction to Flow Cytometry,* Cambridge University Press, New York, 1991, 150–158.
68. Loken, M.R., Immunofluorescence techniques, in *Flow Cytometry and Sorting,* 2nd ed., Melamed, M.R., T. Lindmo, and M.L. Mendelsohn, Eds., Wiley-Liss, New York, 347.
69. Kim, Y.R. and L. Ornstein, Isovolumetric sphering of erythrocytes for more accurate and precise cell volume measurement by flow cytometry, *Cytometry,* 4, 419–427, 1983.

Chapter 10

Modern Techniques for Cell Labeling in Avian and Murine Embryos

Diana K. Darnell and Gary C. Schoenwolf

Contents

0-8493-3342-3/97/$0.00+$.50

or older embryos are being cultured *in vitro*, whole-egg/agar plates should be used (Packard culture).[8,9]

Preparation of Spratt and New Culture Plates

1. Adjust a water bath to 49°C.
2. Open the blunt end of an egg with dull forceps (e.g., Semken forceps* or old watchmakers forceps* that have been filed down after the tips have become damaged); remove thick albumen and chalazae with dull forceps and discard.
3. Collect thin albumen into a graduated cylinder (e.g., 60 ml for 40 plates), then transfer to a sterile 250-ml beaker and place into the 49°C water bath.
4. Make a 0.6% solution of Bacto-Agar™*[b] in saline (e.g., 0.36 g in 60 ml for 40 plates). On a magnetic stirrer/heater, in a 250-ml beaker, mix and bring to a slow boil to dissolve agar. Place the beaker into the 49°C water bath to cool.
5. When the temperatures of both agar/saline and albumen have equilibrated at 49°C, add albumen to sterile agar/saline while stirring vigorously until almost frothy. Return beaker to the water bath and pipette about 2.5 ml agar-albumen into each culture dish* (35 × 10 mm) to produce a confluent layer on the bottom. Avoid transferring bubbles.
6. Allow the culture plates to sit undisturbed at room temperature until culture medium solidifies, then place plates into airtight plastic storage containers lined with wet paper towels. Dishes may be refrigerated for up to 2 to 3 weeks but are best when used fresh.
7. On the day of an experiment, place culture plates in a humidified chamber (e.g., a 15-cm culture dish* with a wet piece of filter paper in the bottom) in a 38°C plate incubator before use. If culture plates will be used for Spratt culture, they can be flooded with sterile saline at this time.

Preparation of Packard Culture Plates

1. Adjust a water bath to 47°C.
2. Homogenize the contents of three cold, unincubated, fertile chicken eggs in the cooled, sterile container of a Waring™*[c] blender.
3. Pour the foamy homogenate into sterile, 50-ml centrifuge tubes* and spin at 14,900 g for 30 minutes at 5°C.
4. Prepare a 50-ml aqueous solution of 6% Bacto-agar,* autoclave for 8 minutes (120°C and 103.5 kPA), and place into a 47°C water bath.
5. Pour supernatant from eggs into a sterile 250-ml glass beaker and warm to 47°C in the water bath.

---

[b] Bacto-Agar is a registered trademark of Difco Laboratories; Detroit, Michigan.

[c] Waring is a registered trademark of Dynamics Corporation of America; New Hartford, Connecticut.

6. When the temperatures of both solutions have equilibrated at 47°C, add the agar slowly to the egg, swirling the beaker without forming bubbles, to a final ratio of 1 part agar to 3 parts egg.
7. Pour 2 to 3 ml of this solution into the center of a warmed culture dish* (60 × 15 mm), swirl gently to coat, and pour off excess liquid to leave a thin (1 to 1.5 mm) coating on the bottom of the dish.
8. Allow the culture plates to sit undisturbed at room temperature until culture medium solidifies, then place plates into airtight plastic storage containers lined with wet paper towels. Dishes may be refrigerated for up to 2 to 3 weeks but are best when used fresh.

2. Spratt and Packard Whole-Embryo Culture

Spratt and Packard cultures involve culturing embryos without the vitelline membranes. This type of culture is ideal for certain kinds of embryo manipulations, including extirpations and transections, and for short term culture of donor embryos prior to collecting tissue for grafting into host embryos. Embryos cultured in Spratt culture develop relatively normally for 24 hours; however, their neuraxis does not extend to the normal lengths seen *in ovo* and in New culture. Cutting off the very caudal end of the area pellucida facilitates axis extension in Spratt culture (see step 13 below). Optimal stages for starting Spratt cultures are from HH mid-stage 3 (stage 3c)[5] to 9; embryos can develop well for up to 24 hours or to HH stage 14 in this culture system. In addition to whole-embryo culture, fragments of embryos can be cultured in isolation using a similar protocol. For older embryos (up to 3 days or HH stage 18), a Spratt-like culture, Packard culture[8,9] can be done using plates containing yolk. Packard culture also works best for blastoderms obtained from unincubated eggs. To culture embryos without the vitelline membranes:

1. Incubate eggs in a humidified, forced-draft egg incubator* until desired stages are reached (see Hamburger and Hamilton[1] for approximate timing).
2. Open the blunt end of an egg with a pair of dull forceps (e.g., Semken forceps* or old watchmakers forceps* that have been filed down after the tips have become damaged) and gently remove thick and thin albumen and chalazae with forceps and by pouring. Try not to puncture the yolk.
3. Pour the remaining yolk, with the blastoderm and attached vitelline membranes, into a sterile dish (we use a Pyrex-brand™* [d] evaporating dish) that is roughly half full of sterile saline.
4. Remove any excess albumen surrounding the yolk using forceps or a pipette. This step is crucial for New culture and makes Spratt and Packard culture easier as well.
5. Orient the yolk with the blastoderm uppermost, then cut around the equator of the yolk with scissors to isolate the blastoderm and vitelline membranes from the yolk.

---

[d] Pyrex is a registered trademark of Corning, Inc.; Corning, New York.

6. Remove the vitelline membranes and attached blastoderm from the yolk by gently lifting up on the cut edge of the vitelline membranes and then gently pealing the vitelline membranes with attached blastoderm away from the yolk.
7. Transfer blastoderm and vitelline membranes with a spoon into a glass petri dish containing sterile saline, and place this dish on the microscope stage.
8. Remove the remaining yolk by puffing saline carefully from a pipette onto the embryo and then aspirating the yolk from the saline, rather than aspirating the yolk directly off of the embryo. With practice, embryos can be collected virtually without adherent yolk.

Note: *For New culture, go to step 1 of Protocol Section I.A.3. For Spratt or Packard culture, continue with step 9.*

9. Remove the blastoderm from the vitelline membranes. To do this, place the blastoderm ventral-side up and find the outer margin where the blastoderm and vitelline membranes are attached. Loosen the margin with forceps. Pipette saline between the blastoderm and vitelline membranes to detach the blastoderm, or detach by gently pulling the two layers apart with forceps.
10. Stage the embryo accurately and record this information.[1-5]
11. Transfer the blastoderm via a wide-mouthed pipette (break off the tip of a Pasteur pipette, cover the broken end with a pipette bulb, and use the wide end for pipetting) to a culture dish using sufficient saline to facilitate orientation of the embryo on the plate.
12. Orient the blastoderm with the desired side up. The ventral side was toward the yolk and will still have some yolk attached; the dorsal side was toward the vitelline membranes and has a smoother surface. Pipette off excess saline to flatten the blastoderm onto the substrate.
13. Trim away the outer two thirds of the area opaca from the blastoderm using a cactus needle-knife (a cactus needle attached to a small dowel), a pulled glass micropipette, or an iris knife. At the caudal end of the blastoderm, remove the area opaca entirely and a little of the area pellucida to ensure good extension of the axis.
14. Add saline dropwise to the culture dish, move the blastoderm to a fresh area of the substrate, and pipette away excess saline and pieces of extirpated area opaca.
15. As desired, make injections (Section I.B.1) or fluorescent-labeled grafts (Section I.B.2) or chimeras (Section I.B.3) or label embryos with retroviruses (Section I.B.4).
16. Place culture plates into a humidified chamber (e.g., a 15-cm culture dish* with wet filter paper in the bottom) in a humidified plate incubator* at 38°C.
17. After the desired period in culture, assess stage,[1,2] evaluate development, fix embryos, and proceed to immunocytochemical labeling (Section I.B.5), *in situ* labeling (Section I.B.6), or paraffin embedding (Section I.C.2).

3. New Whole-Embryo Culture

New[6] culture is ideal for experiments involving grafting from one embryo into another, for microinjection, or for time-lapse videotaping, and where superior development is required from cultured embryos during the first 2 days of incubation. Blastoderm expansion occurs vigorously in New culture. Optimal stages for starting New cultures are between HH stages 1 through 9, with the maximum time of development in culture without degeneration of some tissues equaling 24 hours or HH stage 14, whichever comes first. Packard culture plates (Section I.A.1, *Preparation of Packard Culture Plates*) should be used for HH stages 1 to 2 embryos, and Spratt/New culture plates (Section I.A.1, *Preparation of Spratt and New Culture Plates*) should be used for HH stages 3 to 9 to optimize development. A variation on the New culture method for use with embryo fragments and transections can be found in recent papers by Yuan and co-workers.[10,11] To culture embryos with the vitelline membranes:

1. Follow steps 1 through 8 as described in the previous protocol (Section I.A.2).
2. Sterilize glass or plastic rings*[e] in 70% ethanol; transfer them to a bowl containing sterile saline until ready to use.
3. Orient the blastoderm ventral-side up; that is, with the embryo on top of the vitelline membranes.
4. Gently place a ring on top of the vitelline membranes, with the embryo near the center of the ring. Remove saline from outside the ring until the top of the ring is exposed.
5. Use forceps to lift the part of the vitelline membranes remaining outside the ring over the top of the ring. Do this until the entire edge of the vitelline membranes is lifted up over the top of the ring, making a small bowl. Gently pull the vitelline membranes (as necessary) so that they are taut and the blastoderm is roughly centered in the ring. Then carefully remove saline from the interior of the ring with a pipette.
6. Lift the ring with the attached vitelline membranes and blastoderm using forceps, and transfer it to a culture plate prepared as described in Section I.A.1.
7. Place the culture plate into a humidified chamber (e.g., a 15-cm culture dish* with wet filter paper in the bottom) in a humidified plate incubator* at 38°C.

Note: *For normal development to occur, it is necessary for the blastoderm to remain attached to the vitelline membranes around its circumference. Experiments can be done individually as each embryo is collected, or a group of embryos can be collected and then all of the experiments can be done in a block.*

***In Ovo* Culture.** *In ovo* culture is useful if development beyond day 2 is required. Although normal development can be achieved using this technique for embryos windowed at 2 or more days of incubation, as many as 70% of surviving embryos windowed on day 1 will have some abnormality, usually dysraphic (open) neural tubes, unless steps are taken to minimize these effects.

[e] These rings are not available commercially. They can be cut from glass or plastic tubing by a machine shop with a diamond saw. Dimensions should be approximately 20 mm in diameter, 2.5 mm tall, with a wall thickness of 2 mm.

(In any case, approximately 50% of the embryos will die after windowing.) Here we describe the improved protocol for windowing day-1 eggs which results in fewer than 5% of the surviving embryos developing abnormally in culture.[12] For this protocol, eggs should be incubated with their long axis parallel to the horizon for at least 24 hours. Mark the side facing up during culture so that it can be identified when the window is cut. Eggs for experiments beginning at later stages may be windowed on day 2 and returned to the incubator until needed. Windowing beyond day 2 becomes more difficult because of the development of highly vascular extra-embryonic membranes, which lie just beneath the shell and its membranes. To culture embryos *in ovo:*

1. Eggs to be windowed should be cleaned with a towel damp with 70% ethanol. Begin by puncturing the air space at the blunt end of the egg by piercing the blunt end with a needle. If desired, 1 to 2 ml of thin albumen can be removed from the pointed end of the egg with a syringe (seal the hole with tape after withdrawing the needle). Puncturing the air space and/or removing some thin albumen allows the blastoderm to drop away from the shell and its membranes at the time of windowing, thereby reducing the likelihood that the embryo or its membranes will be damaged during windowing. Windowing involves carefully removing approximately a 1-cm^2 area of egg shell and the underlying shell membranes using a dental drill and forceps.[13] The window is made on the side of the egg that was uppermost during incubation (i.e., over the blastoderm).
2. Embryos can be better visualized for staging and manipulation if a drop of a 1% aqueous solution of a vital stain (neutral red or Nile blue) is applied to the vitelline membranes overlying the blastoderm. Or, vital stains can be more focally applied (our method of preference) on chips of agar impregnated with stain. To do this, a clean, sterile, glass slide is coated on one side with 1% agar. After drying overnight, the agar slides are soaked for a week in 1% neutral red or Nile blue, rinsed with sterile distilled water, and allowed to dry. To stain embryos, place a drop of sterile distilled water or saline on the coated side of the slide and remove a tiny chip of agar from the moistened area using a scalpel blade. Transfer the chip with dull watchmakers forceps* to the vitelline membranes over the embryo, leave it for a brief period of time (usually less than 1 minute) for staining to occur, and remove it with forceps. Do not over stain, as vital stains can be toxic in excess. An alternative method is to inject 0.05% Nile blue in saline or full strength India ink* into the subgerminal cavity (between the embryo and the yolk). India ink is highly toxic, however, prior to day 2 of incubation.
3. To gain access to the embryo, the egg should be cradled in a stable holder (such as a portion of an egg carton) with the window up and light sources directed into the window (fiber optics work best). The vitelline membranes can be torn open with a tungsten needle over the area you wish to manipulate. Dyes or retroviruses can then be injected, or tissue can be removed and donor tissue grafted into the host.
4. After surgical manipulation or injection, sterile saline should be added to bring the embryo up to the level of the window. Then, the window in the shell should be sealed well with tape (Scotch™*[f] brand Super 88 vinyl electrical tape works

[f] Scotch is a registered trademark of 3M Corporation; St. Paul, Minnesota.

best and does not leak) and the egg should be rotated 180° along its long axis so that the embryo is subjacent to undisturbed shell; eggs are then returned to the incubator for the duration of the culture period, sealed window-side down. The addition of saline and rotation can be delayed for up to 3 hours without having a negative impact on development. This is recommended in the case of retroviral labeling so that the virus is not diluted immediately after placement on the embryo (Section I.B.4).

B. Labeling Cultured Embryos

Chick embryos can be labeled with a variety of techniques. Small groups of cells can be labeled with microinjected fluorescent dyes; grafts can be fluorescently labeled or taken from a compatible species and followed by a species-specific marker; the progeny of randomly infected, individual cells can be labeled with retrovirus expressing a reporter gene; or tissue (or position-specific) protein or RNA can be detected by immunocytochemistry or *in situ* hybridization, respectively. Theoretically, single cells can be labeled in chick embryos by iontophoretic injection, and clones of cells derived from the injected cells subsequently can be followed. This technique is described in Section II.B for mouse, where it has been most effectively demonstrated.

1. Extracellular Injection and Multiple Cell Labeling

To trace the movements of small groups of cells, focal injections of fluorescent dye can be made into a specific region of the embryo, and the fluorescent label subsequently can be followed during culture (Figure 10.1). Possible dyes include: CFSE* (5-[and -6]-carboxyfluorescein diacetate, succinimidyl ester), a label that passes through the cell membrane, binds to protein amines in the cytosol, and becomes fluorescent after reaction with nonspecific esterases within the cytoplasm;[14] DiA* (4-[4-dihexadecylaminostyryl]-N-methylpyridinium iodide), DiI* (1,1′-dioctadecyl-3, 3, 3′, 3′-tetramethylindocarbocyanine perchlorate), and DiO* (3,3′-dioctadecyloxacarbocyanine perchlorate), cell membrane labels;[15,16] and Mitotracker Green FM™*[g] and Rhodamine-123,™*[h] mitochondrial labels.[17,18] Each of these dyes can be injected extracellularly as a small bolus, after which a localized population of cells in the immediate vicinity of the injection site becomes labeled. In our hands the cell surface labels give the same labeling pattern after culture as do labels taken into the cells. A mixture of these labels can be injected to take advantage of their different properties. A colored label (e.g., DiI) can be included so that the actual injection can be visually monitored at the time that it is made using conventional light microscopy. Surface (e.g., DiI) and cytoplasmic (e.g., CFSE) labels, visible with different filter sets, can be mixed for internal verification of the experimental results. A fluorescein-containing label (e.g.,

[g] MitoTracker is a registered trademark of Molecular Probes, Inc.; Eugene, Oregon.

[h] Rhodamine 123 is a registered trademark of Molecular Probes, Inc.; Eugene, Oregon.

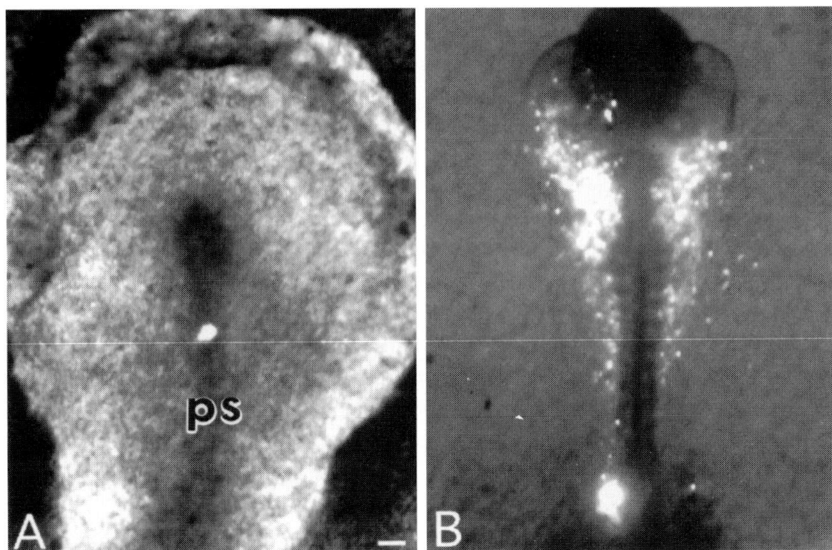

FIGURE 10.1
(**A**) A stage 3 chick embryo that has been injected with fluorescent dye (Rhodamine 123*) near the mid-primitive streak level (white dot in structure labeled ps). (**B**) After 24 hours of culture, fluorescent-labeled cells (white) have migrated into the prospective heart mesoderm and lateral plate mesoderm. The embryo is now at stage 9. Bar = 100 μm.

CFSE) can be included in the mixture so that the embryos can be immunocytochemically labeled after culture (Section I.B.5, *Antibody Labeling of Whole Mounts*) using an anti-fluorescein antibody.* This provides a permanent label that can be detected in whole mounts or in sections after processing for paraffin histology (Section I.C.2).[19] (This technique is superior to photo conversion, which had been used previously to deposit an insoluble marker after fluorescent labeling.) We routinely use a 1:1 solution of DiI:CFSE made separately (2.5 mg of either dye dissolved in DMSO and mixed with 1 ml of 100% ethanol) and mixed at the time of injection. To inject fluorescent dyes:

1. Isolate embryos and prepare for culture on plates.
2. Pull a microinjection pipette from capillary tubing* with an Omega dot fiber™[i] with a 10 to 12-μm tip. We use a Narishige micropipette puller* set for double pull. Optimal settings should be determined empirically for each puller.
3. Backfill the microinjection pipette with a dye solution using a Microfil™*[j] 28 AWG needle (the Omega dot fiber facilitates backfilling).
4. Place the micropipette into the microinjection holder (we use a Narishige hydraulic micromanipulator*) attached to a Picospritzer II™*[k] or similar delivery

[i] Omega dot fiber is a registered trademark of Fredrick Haer & Co.; Brunswick, Maine.
[j] Microfil is a registered trademark of World Precision Instruments, Inc.; Sarasota Florida.
[k] Picospritzer II is a registered trademark of General Valve Corporation; Fairfield, New Jersey.

device for pressure-driven injection. Injection pressure and duration should be determined empirically for each needle.

5. Test the needle by making a practice injection into a dish of sterile saline. One "spritz" should deliver a small but visible bolus of dye if a colored dye is present (e.g., DiI).
6. Manually lower the microinjection pipette until it is near the surface of the embryo (or use a coarse micromanipulator). Lower it the remaining distance using the fine (hydraulic) micromanipulator.
7. Pierce the tissue and use the Picospritzer to deliver some dye. Sometimes the Picospritzer must be activated several times to deliver the desired amount of dye into the tissue. It is better to deliver too little dye per injection than too much; in the latter case, embryos should be discarded. It should be possible to use one pipette to inject several embryos. (If the pipette gets clogged with debris, it sometimes can be cleared by withdrawing the needle from the embryo and wiping the tip with another injection pipette or by temporarily increasing the injection pressure or the duration of the air pulse and activating the Picospritzer a few times.)
8. Withdraw the pipette, record the injection location and size using a fluorescence microscope fitted with an intensified videocamera (we use a Dage-MTI™*[1] camera), image processor (we use a Dage-MTI image processor*), and VCR. DiI, DiA, Mitotracker, and Rhodamine 123 can be viewed using a rhodamine filter set; CFSE and DiO using a fluorescein filter set. Place the embryo back into the incubator and continue with the next embryo.

2. Fluorescent Labeling of Grafts

Another way to label cells using fluorescent dyes is first to soak the entire donor embryo in the dye and then to isolate, wash, and transfer the labeled graft to an unlabeled host embryo (Figure 10.2). This section describes the technique for labeling the graft tissue. CFSE,* Mitotracker Green FM,* Rhodamine-123,* or a mixture of the above (see Section I.B.1), can be used for labeling with this method. The procedure for making chimeras is described in the next section (I.B.3). To label a graft plug:

1. Make up stock and working solutions of the dyes you plan to use. Make a 1% stock solution of CFSE or Rhodamine 123 by mixing 10 mg of CFSE or Rhodamine 123 with 1 ml of DMSO.* Dilute 25 to 50 μl of this stock solution with 10 ml of saline to make a 1× working solution. For a comparable Mitotracker stock solution, mix 50 μg in 74 μl of DMSO; for a 1× working solution, mix 30 μl stock with 2 ml of saline. If mixtures of dyes are desired, make 2× working solutions and mix them 1:1.
2. Set up donor embryos for New (Section I.A.3) cultures.
3. Stain the donor embryo by dropping an aliquot of working solution of one of the three dyes (or a mixture of dyes) onto the embryo and culturing it for 1 to 2 hours at 38°C. Then, wash this donor embryo with saline to remove excess

[1] Dage-MTI is a registered trademark of Dage-MTI, Inc.; Michigan City, Indiana.

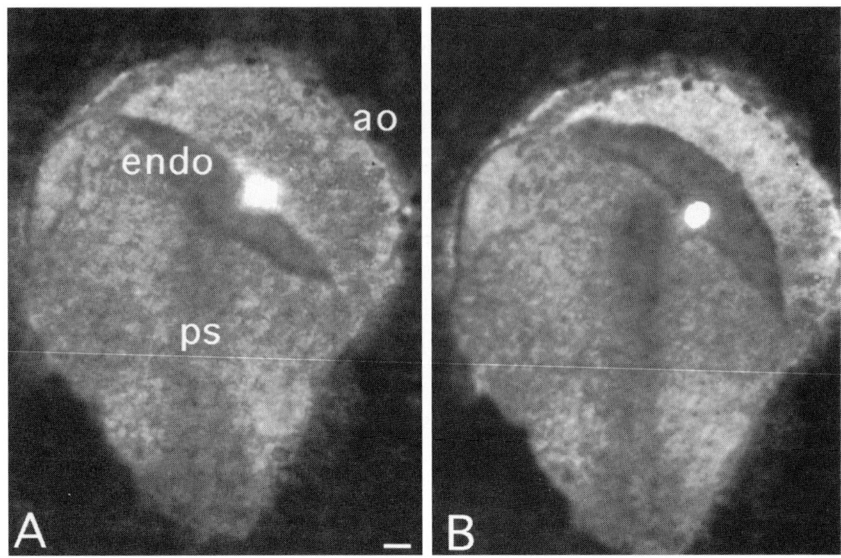

FIGURE 10.2
(A) A stage 3 host embryo has had its endoderm/hypoblast (endo) reflected back from the edge of the area opaca (ao), a hole has been punched in the epiblast adjacent to the primitive streak (ps), and a fluorescently labeled donor plug (white) has been grafted into the hole. **(B)** The endoderm/hypoblast has been folded back over the graft. Bar = 100 μm.

dye, isolate the graft, and transfer it to a depression slide* for additional washing with saline.

4. Transfer the graft to the host as described in the next section (Section I.B.3).
5. Examine embryos at time zero and after culture using a fluorescence microscope.
6. If a fluorescein-containing dye was used, immunolabel grafted cells when cultures are terminated using an anti-fluorescein antibody* and whole-mount immunocytochemistry (Section I.B.5, *Antibody Labeling of Whole Mounts*) to permanently label the graft cells and allow you to see the graft cells in whole mount and sections (Section I.C.2) using conventional light microscopy.

3. Making Chimeras

Another way to label avian embryos is to transplant tissue between two embryos to form a chimera. This can be done between embryos of the same stage (isochronic) or different stages (heterochronic); between similar tissues or regions (homotopic) or different tissues or regions (heterotopic); and between tissues of the same species (intraspecific) or different species (interspecific). The classic technique for using chimeras to study development involves grafting quail tissue into chick embryos (interpretation of these experiments is based on the assumption that grafts between these two similar species are equivalent to grafts from intraspecific chimeras) and identifying the grafted cells after sectioning using the Feulgen-Rossenbeck histological stain.[20] With

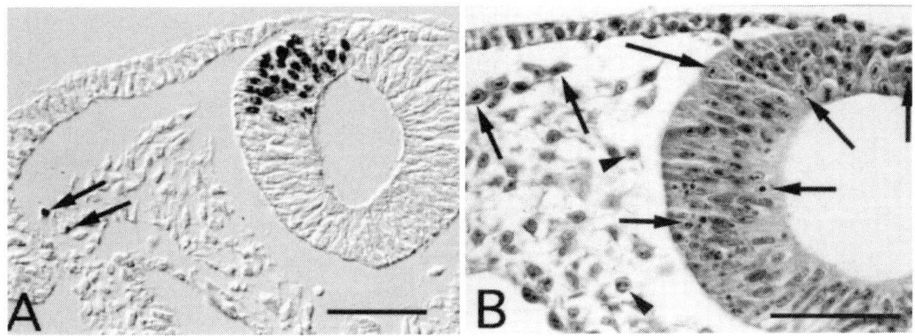

FIGURE 10.3
(A) A graft (similar in placement to that shown in Figure 10.2, but from a quail donor into a chick host) was immunolabeled after 24 hours in culture with an anti-quail antibody and processed for paraffin sectioning (7 μm thick). Quail cells (almost exclusively in the neural tube) are visible due to their black nuclei. *Note:* single quail cells in the mesenchyme (arrows) can be detected easily. **(B)** A similar chimera stained with the Feulgen-Rossenbeck histological stain after paraffin sectioning reveals its graft in the neural tube and neural crest (dark nucleolar staining; arrows). Scattered individual graft cells within the mesenchyme can be difficult to detect (arrowheads). Bars = 100 μm.

this technique, groups of quail cells can be identified by the characteristic appearance of their nucleolus.[21] The drawback to this technique is that some chick cells also appear to have a similar nucleolar characteristic and some quail cells do not have the characteristic, making the exact identification of individual cells as chick or quail problematic. We find that the anti-quail antibody* (QCPN) used with the immunocytochemistry protocol (Section I.B.5) makes *every* quail cell easily discernible from the neighboring chick cells and thus is preferable to the histological stain (Figure 10.3). However, because the Feulgen-Rossenbeck procedure may be useful in some experiments involving multiple labels,[22, 23] it is included (Section I.B.7).

Labeling embryos by grafting can be used experimentally to test several developmental changes. Chimeras can be used to ascertain the fate of a given region of the blastoderm (all layers of the embryo) or of a single layer (e.g., the epiblast) by grafting isochronic, homotopic grafts.[5, 23] They can be used to determine the developmental potential or state of commitment of a group of grafted cells by grafting heterochronic or heterotopic grafts.[24-32] Conservation of cell signaling molecules can be assessed in interspecific chimeras.[33]

Typically, quail embryos in Spratt culture (Section I.A.2) have been used as donor embryos for chimeras (except when fluorescently labeled grafts are used), and chick embryos in New culture (Section I.A.3) have been used as hosts. This is due solely to the larger size of the chick embryo, which facilitates the delicate manipulations involved in grafting. Quail embryos also can be New cultured and used as hosts, but smaller, 15-mm diameter rings are required. With the advent of fluorescently labeled grafts (Section I.B.2), chick embryos can be used as donors and chick grafts can be grafted to chick hosts. To make chimeras:

1. Prepare culture plates for Spratt and New culture (Section I.A.1).
2. Put donor embryos at an appropriate stage into Spratt culture (Section I.A.2) and host embryos at an appropriate age into New culture (Section I.A.3).
3. Using a needle (a pulled glass microelectrode, a tungsten needle, or a cactus needle attached to a small dowel), cut a fragment out of the donor blastoderm. Or, if circular plugs are desired (e.g., to facilitate healing of the graft into the epiblast), pull a glass pipette, score it with a diamond pencil, and break off the tip to the desired diameter (e.g., 50 to 200 μm). Select pipettes with straight tips, and polish these on abrasive film* to obtain an even edge. Use these on the blastoderm like a hole punch. To obtain a single-layered graft, cut through the entire blastoderm to isolate the graft and then separate the layers of the graft with a brief treatment in collagenase* (1000 to 2000 IU/ml in Tris buffer, pH 7.1), dispase* (0.24 U/ml), or trypsin*/EDTA/glycine[m] (0.4%). Rinse in several changes of saline to stop the reaction and to clear away the enzyme.
4. Remove a similar size fragment or plug from the host. Single layers can be removed from the host by careful dissection. For single-layer grafts into the epiblast, add saline to the cultured host embryo until the embryo is submerged. Cut through the hypoblast/endoderm just peripheral to the boundary between the area pellucida and area opaca, and carefully reflect back the hypoblast/endoderm (use an eyebrow hair attached with nail polish to a dowel as a manipulator). Remove the region from the host epiblast that will be replaced by the graft.
5. Transfer the graft piece between culture plates using a small pipette and manipulate the graft into the hole in the host. In the case of an epiblast graft, replace the hypoblast/endoderm by "rolling it" back across the graft, then remove the excess saline. Grafts can also be inserted into the mesodermal layer either by first retracting the endoderm, placing the graft on the mesoderm, and then covering the graft with endoderm or by creating a tunnel between the endoderm and epiblast with an eyebrow hair tool and pushing the graft into the tunnel to the desired location. This technique allows for more control over graft location, as the graft is less likely to float around than when the endoderm has been reflected back.
6. Culture the embryos in a humidified chamber at 38°C until the desired stage is reached, for up to 24 hours.
7. After culture, quail-chick chimeras can be fixed and treated for whole-mount immunocytochemistry using the anti-quail antibody* (Section I.B.5, *Antibody Labeling of Whole Mounts*) to detect graft cells in whole mount. After immunocytochemistry, embryos can be sectioned (Section I.C.2) and individual quail cells can be detected in sections.

4. Retroviral Labeling

Retroviral labeling of cells in the early embryo (Figure 10.4) involves infection of one to many individual cells with a retrovirus that has been engineered to

[m] To make trypsin/EDTA/glycine, mix 0.4 g trypsin, 0.4 g EDTA (ethylenediaminetetraacetic acid), and 7.5 g glycine with 60 ml distilled water; stir for 1 hour. Add 10 *N* NaOH dropwise until the EDTA goes into solution, then 2 *N* HCl dropwise until the pH is 7.2. Bring the volume to 100 ml with distilled water.

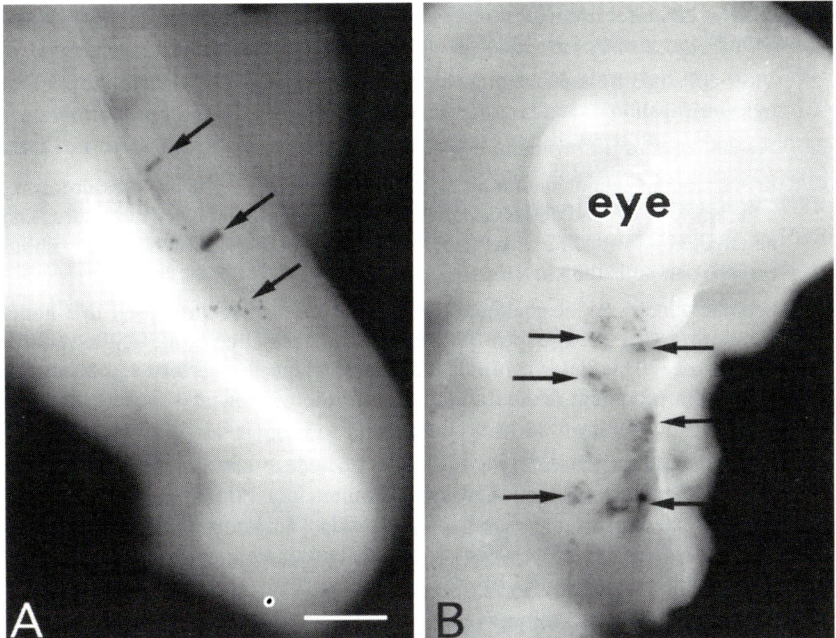

FIGURE 10.4
Two examples of retroviral infections of early chick embryos. Infected cells are dark (arrows), and both small and large groups of cells can be seen. **(A)** Infected cells are principally in the neural tube. **(B)** Infected cells are principally in the branchial arches (labeled cells derived from the neural crest). Bar = 100 μm.

be replication incompetent and to contain a reporter gene such as *lac-Z* (for beta-galactosidase).[34-41] The progeny of these cells can be followed after cell division, because the viral DNA (including the marker) is replicated with the cellular DNA and, therefore, is not diluted or lost. In this way, cell lineage can be followed throughout development. Drawbacks to this technique include: (1) serious difficulty in titering the virus to get an appropriate (small) number of infections per embryo so that individual clones can be distinguished from one another; (2) problems distinguishing progeny of multiple infections after significant cell movement has occurred; (3) differential infection rates for different tissue types (high for cardiac and neural crest cells, low for endoderm), making it impossible to use retroviral infection to measure the number of precursor cells in different lineages; (4) labeling of only dividing cells; (5) potential irregular results due to knock-outs of essential genes during integration of the virus, which could result in abnormal development of the progeny of infected cells; and (6) potential human risk of infection from some viruses. Using retroviruses to mark avian cells during gastrulation and neurulation has not worked well for us; however, it has worked for others with long-term incubation and has proven useful under some circumstances. Retroviral vectors can also be used to label murine cells for lineage analysis.[40] To use retroviral vectors to label avian cells:

5. Carefully wipe excess fluid off of the slides and add a 1:200 dilution of horse-anti-mouse* (HAM) secondary antibody. Incubate the slides in the humidified chamber for 1 hour.
6. While there are still 15 minutes left in the above incubation, make up the ABC Complex using the Vectastain™*[p] ABC kit. The ABC complex must sit at room temperature for 30 minutes before it is used. Add Reagent A to PBS at a ratio of 1:200 (15 μl to 3 ml of PBS). Shake the solution to mix it. Add Reagent B to the solution at a ratio of 1:200 (15 μl).
7. After the HAM incubation, rinse slides off with PBS and put them through three 5-minute washes of PBS.
8. Wipe the edges of the slides and add 100 μl of the ABC complex to the sections. Incubate the slides in the chamber for 30 minutes at room temperature.
9. Rinse slides with PBS and wash them three times for 5 minutes each with PBS.
10. During the last 5-minute wash, prepare the DAB (diaminobenzadine) solution using Sigma Fast™*[q] DAB* tablets. *Note:* DAB is a carcinogen and appropriate caution should be taken (e.g., wear gloves during the procedure, wash DAB-contaminated items with bleach, and dispose of the DAB appropriately).
11. After the last 5-minute PBS rinse, wash slides in distilled H_2O for 2 minutes. Wipe edges of slides and add 100 μl of DAB solution to the sections. Leave the DAB on the sections for approximately 9 minutes; however, monitor the reaction the entire time. If the sections appear to be turning brown too quickly, you may want to stop the reaction early.
12. After the reaction is complete, rinse the slides with PBS, collecting the wash in an appropriate DAB waste container. Put the slides into water to stop the reaction.
13. Dehydrate the slides for 2 minutes each through the following ethanol series: 35, 50, 70, 70, 95, 95, 100, and 100%, then two times for 3 minutes each in Histosol.
14. Add cover slips to slides using Pro-Texx™*[r] mounting medium.

Antibody Labeling of Whole Mounts

As with immunocytochemistry on sections, the point of this technique is to localize a specific protein with an antibody. We do almost all of our immunocytochemistry on whole embryos, then section (Figure 10.5). It is simple, and the labeling pattern can be seen in the context of the whole embryo, as well as in subsequent sections if desired. This procedure also works after *in situ* hybridization (Section I.B.6) to give double-labeled embryos. For double labeling, we do not intensify the DAB reaction as we routinely do with single labeling (see step 14 below). Non-intensified reactions give good color contrast between the purple *in situ* label and red/brown immunocytochemistry label. Both chick and mouse embryos (as well as other species[46]) work well with this technique, although extra care must be taken with young mouse embryos because they are small and easily lost during pipetting. We do the

[p] Vectastain is a registered trademark of Vector Laboratories; Burlingame, California.

[q] Sigma Fast is a registered trademark of Sigma Chemical Co.; St. Louis, Missouri.

[r] Pro-Texx is a registered trademark of Lerner Laboratories; Pittsburgh, Pennsylvania.

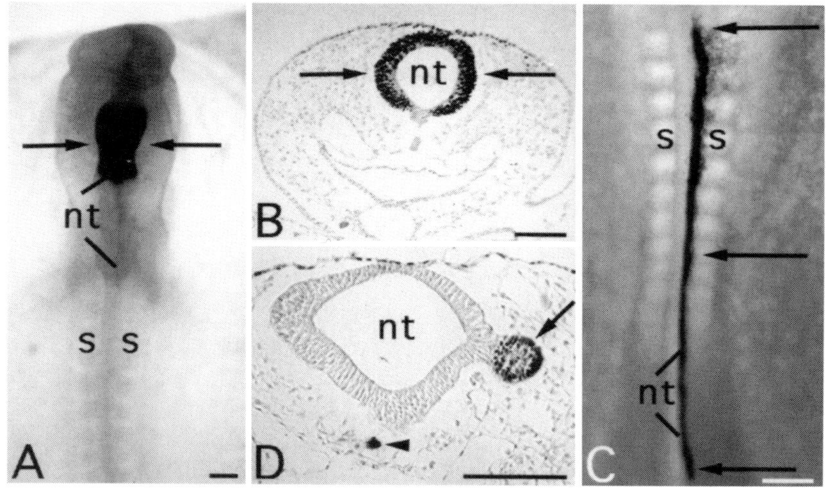

FIGURE 10.5
Immunocytochemical labeling of whole-mount embryos can be used to enhance several distinct experimental paradigms. Normal protein expression (in this case, Engrailed-2 labeled with 4D9; arrows) can be seen in the whole embryo (**A**) and in sections after processing for paraffin histology (**B**). Grafted tissue can be identified in whole mount (**C**) (in this case with an anti-quail antibody) or in sections (**D**) (in this case using a tissue specific antibody because a specific tissue or its precursor cells had been grafted; a notochord has been grafted and an antibody, Not-1, has been used to label it [arrow] as well as the host's notochord [arrowhead]). nt = neural tube; s = somite. 4D9 was originally described by Patel et al.[46] Not-1 was originally described by Yamada et al.[42] Bars = 100 μm.

entire procedure between fixation and DAB reaction in a 96-well plate* (we use Nunclon™[s] microwell plates). Again, specifics for antibodies and serum are examples and appropriate choices must be made for a given experiment. To label whole mounts:

1. Dissect off vitelline membranes and transfer embryos to an appropriate dish for fixation. Fix embryos for 60 minutes in PEM-FA at room temperature. PEM = 0.1 *M* Pipes (1.73g per 50 ml), 2 m*M* EGTA (38.5 mg per 50 ml), and 1 m*M* $MgSo_4$ (12.3 mg per 50 ml). Adjust pH to 6.95 with concentrated HCl; PEM can be stored at 4°C for several weeks. PEM-FA = 9 parts PEM to 1 part 37% formaldehyde. Make within 24 hours of use and refrigerate until needed.
2. Wash embryos two times for 5 minutes each in PBS at room temperature (when doing washes, place on shaker; this may reduce background staining).
3. Wash embryos two times for 5 minutes each in PBT; PBT = PBS (100 ml), 0.2% BSA* (0.2 g), and 0.1% Triton X-100* (100 μl).
4. Wash for 30 minutes in PBT.
5. Incubate for 30 minutes in PBT + N; PBT + N = 4 ml PBT + 200 μl normal goat serum (NGS).* (NGS may be heat inactivated by placing in a water bath at 56°C for 30 minutes, and aliquots should be frozen at –20°C.)

---

[s] Nunclon is a registered trademark of Intermed Nunc; Roskilde, Denmark.

6. Remove PBT + N and replace with supernatant containing primary antibody (e.g., mouse-anti-chicken epitope primary antibody) at an appropriate concentration. Incubate for 4 hours at room temperature or overnight at 4°C. (Longer incubations are good for weakly labeling antibodies. We usually use the overnight incubation for convenience.)
7. Wash three times for 5 minutes each and four times for 30 minutes each with PBT.
8. Incubate for 30 minutes with PBT + N.
9. Replace PBT + N with goat-anti-mouse secondary antibody* (IgG) conjugated to peroxidase (dilution 1:200 in PBT + N).
10. Incubate for 4 hours at room temperature or overnight at 4°C.
11. Wash three times for 5 minutes each and four times for 30 minutes each with PBT.
12. For the peroxidase reaction, add 2 ml of PBT plus 1 ml of DAB solution (1 mg/ml DAB* in PBT). Reaction can be intensified (changes color from red/brown to black) by adding 75 µl 1% CoCl or 60 µl $Ni(NH_4)_2(SO_4)_2$ or both to the original 3 ml DAB-PBT. Set aside 1 ml DAB-PBT for the next step and use 2 ml now. Preincubate for 10 minutes in DAB-PBT.
13. Add 5 µl of a 0.6% solution of fresh H_2O_2 to 1 ml DAB-PBT in a watch-glass, or other shallow container immediately before adding the embryos. Place the embryos in this solution a few at a time and allow the reaction to proceed while examining them under the microscope to ensure that they label to the desired intensity. Adjust the H_2O_2 concentration if the reaction takes less than 2 minutes or more than 6 minutes.
14. Return embryos to PBS in a fresh well of the 96-well plate and wash several times with PBS to stop the reaction.

6. *In Situ* Hybridization: Whole Mount

Whole-mount *in situ* hybridization takes advantage of the wide variety of currently available cDNAs cloned from genes expressed in specific cell types or regions of early embryos and from which probes for these cells or regions can be made. There are several protocols available for this technique. We have successfully used a protocol by Wilkinson,[47] with slight modifications, for chick embryos younger than day 3 and for mouse embryos from day 7.5 to 10.5 (Figure 10.6), although later stages will likely also work. The Wilkinson protocol is described briefly here and our modifications are mentioned. For the full protocol, please refer to the original reference. Double labeling with two *in situ* probes has been reported,[48,49] but it seems difficult unless both signals and probes are strong. Double labeling with *in situ* hybridization and whole-mount immunocytochemistry (Section I.B.5, *Antibody Labeling of Whole Mounts*) works well with this protocol with the minor modifications mentioned below. To use *in situ* hybridization on whole mounts:

1. Embryos should be collected and fixed in 4% paraformaldehyde in PBS with 0.1% Tween-20 (PT) and rinsed in PT; blind pouches such as the prosencephalon and heart should be opened to prevent label from becoming trapped and giving

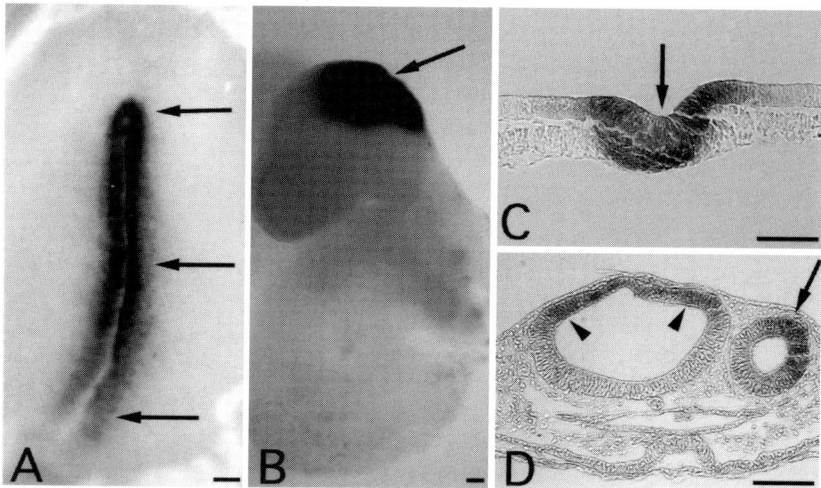

FIGURE 10.6
Whole-mount *in situ* hybridization works well at a variety of stages and in both chick and mouse embryos. **(A)** A stage 3 chick embryo labeled in the primitive streak with a riboprobe for Brachyury (arrows). **(B)** A day 9 mouse embryo labeled at the midbrain-hindbrain junction with a riboprobe for *Engrailed-2* (arrow). **(C)** Labeled embryos can be sectioned to give additional morphological information, as in this cross-section of Hensen's node (arrow) from a stage 6 chick labeled with a *Cnot* riboprobe. **(D)** Gene expression in grafted tissue can also be detected as seen here in grafted neuroectoderm (arrow) expressing *Wnt-1* along with the host neural tube (arrowheads). The Brachyury cDNA was originally described by Kispert et al.[76] The *Engrailed-2* cDNA was originally described by Joyner and Martin.[77] The *Cnot* cDNA was originally described by Stein and Kessel.[78] The chick *Wnt-1* cDNA was obtained from J. McMahon (Harvard University). Bars = 100 μm.

a false-positive result. Some area opaca left on younger avian embryos facilitates seeing them when they are in formamide solutions and become nearly transparent. Embryos are dehydrated through a graded methanol series and stored at –20°C. We put the embryos in flat-bottomed, glass scintillation vials* for the entire procedure.

2. Appropriate sense (negative control) and antisense (experimental) probes can be generated from linearized cDNAs using an appropriate RNA polymerase and digoxigenin-labeled nucleotide mix. Probes can be evaluated on an agarose gel and stored frozen until use, or they can be used fresh.
3. Chick-embryo powder also can be made in advance and works for blocking nonspecific binding sites in both chick and mouse experiments, although ideally the powder should be made from the appropriate species. When enough probes and embryos have been generated to warrant doing the time-consuming *in situ* hybridization procedure (we usually try to do around 12 tubes at a time with multiple embryos in each), then proceed.
4. Day 1 involves inactivating endogenous phosphatases with hydrogen peroxide, increasing permeability with proteinase K (we do not use proteinase K on embryos younger than HH stage 10, using 5 μg/ml for 5 minutes for chick embryos with closed neural tubes and day 7.5 to 8.5 mouse embryos, and up to 10 μg/ml for 15 minutes for day 9.5 to 10.5 mouse embryos), postfixing (can

be eliminated if no proteinase K is used), prehybridizing, and adding the RNA probe. We dilute our probes to 0.25 μg/ml in hybridization mix and replace the prehybridizing mix with 0.5 ml of this — just enough to cover the embryos — and incubate overnight in a 70°C shaking water bath.

5. Day 2 involves washing off the unbound probe, blocking with normal sheep serum* and incubating with the anti-digoxigenin antibody* coupled to alkaline phosphatase. We also use about 0.5 ml/tube for this step, so 3 mg of embryo powder blocks 1 μl of antibody and makes 2 ml final volume for four tubes.
6. Day 3 involves more washing and the final alkaline phosphatase reaction, for which we use BM Purple AP-substrate precipitating,* instead of NBT (nitroblue tetrazolium salt) and BCIP (5-bromo-4-chloro-3-indolyl phosphate). Reactions take from a few minutes to a few hours depending on the strength of the signal.
7. If sections are desired, embryos with strong signals can be processed for paraffin histology, as listed in Section I.C.2. Embryos with weaker signals can be processed similarly, with the exception that two changes of isopropanol (30 minutes) substitute for Histosol (5 minutes) in the two steps prior to embedding, plus one added step of 1:1 isopropanol:paraffin for 30 minutes before the three infiltration steps. Also, embryos are rehydrated, after sectioning and removal of paraffin with Histosol (two changes of 5 minutes each), in a graded series of ethanol solutions (100,100, 95, 95, 70, 50, and 35%), to water, then changed to PBS; 90% glycerol:10% PBS is used as mounting medium for adding cover slips. Photography should be done immediately, because the signal will fade rapidly over time.

7. Feulgen-Rossenbeck Histological Stain

The Feulgen-Rossenbeck stain has been used to differentiate quail cells from chick cells in transplantation chimeras by virtue of its unique pattern of staining of nucleoli in quail cells.[21] This technique allows investigators to detect groups of quail cells, but is not always accurate at the single cell level, as is the anti-quail immunocytochemical label (Figure 10.3). For this reason, we prefer the immunocytochemical labeling procedure (Section I.B.5). However, immunocytochemistry may not always be a viable option due to experimental constraints, so we include the Feulgen-Rossenbeck protocol here. For this procedure, embryos should be fixed in Vakaet's fixative (70% absolute ethanol, 20% formaldehyde [37%], and 10% glacial acetic acid). Strictly aldehyde-based fixatives (e.g., PBS-PFA) may be used, but the timing and concentration of the Schiff's reaction must be reduced, and the nucleolar labeling is less reliable. To stain for quail nucleoli:

1. Prepare Schiff's reagent in advance. We get the best results when we make the Schiff's reagent ourselves, rather than buying a commercially available stock solution. To make the Schiff's reagent, mix 6.0 g pararosanilin hydrochloride,* 9.0 g sodium bisulfite ($NaHSO_3$), 567 ml distilled water, and 6 ml concentrated HCl in a 1-l beaker covered with Parafilm or aluminum foil. Stir for at least 1 hour on a magnetic stirrer. After stirring, place beaker in a dark cabinet and allow it to stand overnight at room temperature. The next day, decolorize by adding 9.0 g decolorizing carbon* and stirring on a magnetic stirrer for

30 minutes; then filter through Whatman™*[t] #1 filter paper. Stain should be clear or straw color. If pink, add more decolorizing carbon, stir thoroughly, and filter as above. At this point, stain is ready for use. Schiff's reagent must be at 23 to 24°C for staining. To store Schiff's reagent between uses, keep it in a tightly closed, brown bottle in the refrigerator. To determine if Schiff's reagent has lost its potency, with a pipette add a few drops of Schiff's reagent to 10 ml of formaldehyde (37 to 40%). A good solution changes color rapidly to reddish purple; however, if the color changes slowly and becomes blue-purple, the reagent is breaking down. (A white precipitate sometimes forms while storing the Schiff's reagent, but this does not seem to affect staining.) If stored properly, Schiff's reagent should be active for at least 6 weeks.

2. Deparaffinize and hydrate sections to water with these changes: 10 minutes Histosol, 5 minutes Histosol, 2 minutes each of ethanol solutions (100, 100, 95, 95, 70, 50, and 35%), and 2 minutes each of two changes of distilled water.
3. Hydrolyze sections in three changes of *freshly prepared* 1*N* HCl: 2 minutes at room temperature, 6 minutes at 60°C, and 2 additional minutes at room temperature; follow this with 1 change (2 minutes) of clean distilled water.
4. Stain in Schiff's reagent for 1 hour at 24°C.
5. Bleach sections in three changes of fresh bisulfite solution, 2 minutes each. Replace with *fresh* bisulfite solution after *every rack* of slides. Bisulfite solution = 4.0 g sodium bisulfite ($NaHSO_3$), 40 ml 1*N* HCl, and 760 ml distilled water; always use fresh bisulfite solution.
6. Rinse in running tap water, 5 minutes. At this point, check sections to see if nuclei are stained bright pink. If not, place slides back into Schiff's reagent, bisulfite solutions, and running tap H_2O as above and then re-check staining of nuclei.
7. Dehydrate sections through a graded series of ethanols up to 90% ethanol (i.e., 35, 50, 70, and 90%), 2 minutes each.
8. Counterstain in 0.1% Fast Green FCF* in 95% ethanol for 30 seconds. (The older the Fast Green, the faster the sections will stain.)
9. Complete the dehydration procedure (i.e., 95, 95, 100, and 100% ethanol; 2 minutes each).
10. Clear in two changes of Histosol* (first change, 2 minutes; second change, 3 minutes).
11. Add cover slips using Pro-Texx* mounting medium and allow to dry.

C. Microscopy and Processing

Most of our surgeries and embryo manipulation are done with the aid of a dissecting microscopes, and embryos are documented with photography or video at time zero and at completion of culture. If continuous visualization of development is desired, embryos can be cultured in a heated environment on a microscope set up for time-lapse videotaping. This technique is especially

---

[t] Whatman is a registered trademark of Whatman Corp.; Clifton, New Jersey.

64. Mintz, B., Experimental recombination of cells in the developing mouse egg. Normal and lethal mutant genotypes, *Am. Zool.*, 2, 541, 1962.
65. Mintz, B., Formation of genotypically mosaic mouse embryos, *Am. Zool.*, 2, 432, 1962.
66. Gardner, R.L., Mouse chimaeras obtained by the injection of cells into the blastocyst, *Nature*, 220, 596, 1968.
67. Gardner, R.L., An investigation of inner cell mass and trophoblast tissues following their isolation from the mouse blastocyst, *J. Embryol Exp. Morphol.*, 28, 279, 1972.
68. Moustafa, L.A. and Brinster, R.L., The fate of transplanted cells in mouse blastocysts *in vitro*, *J. Exp. Zool.*, 181, 181, 1972.
69. Moustafa, L.A. and Brinster, R.L., Induced chimerism by transplanting embryonic cells into mouse blastocysts, *J. Exp. Zool.*, 181, 193, 1972.
70. Brinster, R.L., The effect of cells transferred into the mouse blastocyst on subsequent development, *J. Exp. Med.*, 140, 1049, 1974.
71. Fredrick, G. and Soriano, P., Promoter traps in embryonic stem cells: a genetic screen to identify and mutate developmental genes in mice, *Genes Dev.*, 5, 1513, 1991.
72. Kusakabe, M., Yokoyama, M., Sakakura, T., Nomura, T., Hosick, H.L., and Nishizuka, Y., A novel methodology for analysis of cell distribution in chimeric mouse organs using a strain specific antibody, *J. Cell Biol.*, 107, 257, 1988.
73. Bürki, K., Chimeras, in *Experimental Embryology of the Mouse*, Vol. 19, *Monographs in Developmental Biology*, Sauer, H.W., Ed., Karger, Basel, 1986, chap. 5.
74. Hogan, B., Beddington, R., Costanitini, F., and Lacy, E., Summary of mouse development, in *Manipulating the Mouse Embryo: A Laboratory Manual*, Cold Spring Harbor Laboratory Press, Cold Spring Harbor, NY 1994.
75. McLaren, A., *Mammalian Chimaeras*, Cambridge University Press, Cambridge, 1976.
76. Kispert, A., Ortner, H., Cooke, J., and Herrmann, B.G., The chick Brachyury gene: developmental expression pattern and response to axial induction by localized activin, *Dev. Biol.*, 168, 406, 1995.
77. Joyner, A.L. and Martin, G.R., *En-1* and *En-2*, two mouse genes with sequence homology to the Drosophila *engrailed* gene; expression during embryogenesis, *Genes Dev.*, 1, 29, 1987.
78. Stein, S. and Kessel, M., A homeobox gene involved in node, notochord and neural plate formation in chick embryos, *Mech. Dev.*, 49, 37, 1995.

Appendix A: Information on Supplies, Equipment, and Software

Supplies	Catalog No.	Supplier
Abrasive film	6775E38	Thomas Scientific
Acrodisc filters (0.2 μm)	09-730-218	Fisher (Gelman Sciences)
(0.45 μm)	09-730-219	Fisher (Gelman Sciences)
Anti-digoxigenin antibody	1093-274	Boehringer Mannheim
Anti-fluorescein antibody (IgG, polyclonal)	A-889	Molecular Probes, Inc.
Anti-quail antibody	QCPN	Dev. Studies Hybridoma Bank
Bacto-Agar	0140-01	Difco Laboratories
BM purple AP substrate precipitating	1442-074	Boehringer Mannheim
BSA for PBT (immunocytochemistry)	A-7906	Sigma
BSA for PB1 mouse medium	A-4503	Sigma
Capillary tubing (for holding pipette)	1B100-3	World Precision Instruments
Capillary tubing with Omega Dot Fiber	30-31-1075	FHC
CFSE	C-1157	Molecular Probes, Inc.
Collagenase (type VII, highly purified)	C-8176	Sigma
Culture dishes (35 × 10-mm)	T4155-1	VWR
(60 × 15-mm)	D1990-30	VWR
(15-cm)	08-757-148	Fisher Scientific
DAB (for whole mount)	D-5905	Sigma
Decolorizing carbon	E344-7	Thomas Scientific (Baker)
Depression slides	12-565A	Fisher Scientific
DiA	D-3883	Molecular Probes, Inc.
DiI	D-282	Molecular Probes, Inc.
DiO	D-275	Molecular Probes, Inc.
Dispase	210455	Boehringer Mannheim
Disposable sterilization filter unit (0.2 μm)	09-740-25A	Fisher Scientific
DMF	D8654	Sigma
DMSO	D8779	Sigma
Fast Green FCF	F-99	Fisher Scientific
Fetal bovine serum	230-6140AG	Gibco BRL
Glass scintillation vials	R2550-5	VWR
Goat-anti-mouse secondary antibody	115-035-003	Jackson Immunoresearch Laboratories
Goat-anti-rabbit secondary antibody	605-220	Boerhringer Mannheim
Goat serum (normal)	G-9023	Sigma Immunochemicals
Hanker Yates reagent	08661	Polysciences
Histosol	HS100	National Diagnostics
Horse-anti-mouse secondary antibody	BA2001	Vector Laboratories
Horse serum (normal)	S2000	Vector Laboratories
	H-0146	Sigma

Supplies	Catalog No.	Supplier
Horseradish peroxidase (type IV)	P-8375	Sigma
India ink (fount)	Black	Pelikan
Manomark pens	13-385-1	Fisher Scientific
Mitotracker Green EM	M7514	Molecular Probes
Microfil 28 AWG needle	MF28G-5	World Precision Instruments, Inc.
NP-40 detergent (tergitol)	127087-87-0	Sigma
Paraplast X-tra paraffin	8889-503002	VWR Scientific
Pararosanilin hydrochloride	S903-03	Thomas Scientific (Baker)
Peel-A-Way embedding molds	18985-1	Polysciences
Penicillin/streptomycin	15140-015	Gibco BRL
Polybrene (hexadimethrine bromide)	H9268	Sigma
Potassium ferrocyanide	LC19060-1	Fisher Scientific
Pro-Texx mounting media	M7635-1	VWR
Pyrex-brand evaporating dish	08-710B	Fisher Scientific
Rat Serum (immediately centrifuged)	BT-4095	Harlan Biotechnologies
Rhodamine dextran	D-1817	Molecular Probes
Rhodamine 123	R-302	Molecular Probes
Rings, New culture	—	R&D Products
Scotch brand Super 88 electrical tape	06143	3M
Semken forceps	11008-13	FST
Sheep serum (normal)	S-3772	Sigma Immunochemicals
Sigma Fast DAB (for slides)	D-4293	Sigma
Triton X-100	X-100	Sigma
Trypsin	152-15-9	Difco Laboratories
Tyrode's salt solution	T2397	Sigma
Vectastain ABC kit (Standard Elite)	PK-6100	Vector Laboratories
Watchmaker's forceps	11251-10	FST
Whatman #1 filter paper	09-805G	Fisher Scientific
X-gal	B4252	Sigma
50-ml centrifuge tubes	05-529-1D	Fisher Scientific
50-ml polypropylene Falcon BlueMax tubes	14-959-49A	Fisher Scientific
96-well plates	4616-7008	USA/Scientific Plastics

Equipment and Software	Catalog No.	Supplier
Axovideo	2.0	Axon Instruments
Dage-MTI camera	CCD72	Dage-MTI, Inc.
Dage-MTI image processor	DSP 2000	Dage-MTI, Inc.
Dayton Timewatch garden timer	2E357A	Dayton Electric Manufacturing Co.
Egg incubator	Roll-x	Marsh
Electrometer	51413	Stoelting Co.
Narishige hydraulic micromanipulator	MO-203	Narishige
Narishige micropipette puller	PP-83	Narishige
Optronix camera	DE1-470T	S & M Microscopes, Inc.
Panasonic time-lapse video recorder	RS-232C	Panasonic
Panasonic videocamera	WV-5100HS	Panasonic
Picospritzer II (P/N)	52-100-010	General Valve Corp.
Plate incubator	516D	Fisher Scientific
Waring blender (7011)	31BL92	Waring
Wild dissecting microscope	M8	Wild Heerbrugg
Uniblitz shutter driver/timer	RS-232C	Vincent Assoc.

Appendix B: Addresses of Suppliers

Axon Instruments
1101 Chess Drive
Foster City, CA
phone: 415-571-9400
fax: 415-571-9500

Boehringer Mannheim
9115 Hague Road, P.O. Box 50414
Indianapolis, IN 46250-0414
phone: 800-262-1640

Dage-MTI, Inc.
701 N. Roeske Avenue
Michigan City, IN 46360
phone: 219-872-5514
fax: 219-872-5559

Dayton Electric Manufacturing Co.
5959 W. Howard Street
Chicago, IL 60648
phone: 847-647-0124

Developmental Studies Hybridoma Bank
Department of Biological Sciences
University of Iowa
Iowa City, IA 52242
phone: 319-335-3826
fax: 319-335-2077

Difco Laboratories
P.O. Box 1058A
Detroit, MI 48232
phone: 800-521-0851

Fine Science Tools, Inc.
373-G Vintage Park Drive
Foster City, CA 94404
phone: 800-521-2109
fax: 800-523-2109

Fisher Scientific
711 Forbes Avenue
Pittsburgh, PA 15219-4785
phone: 800-766-7000
fax: 800-926-1166

Frederick Haer and Company (FHC)
P.O. Box 337
Brunswick, ME 04011
phone: 800-326-2905
fax: 207-729-1603

General Valve Corp.
19 Gloria Lane, P.O. Box 1333
Fairfield, NJ 07004
phone: 201-575-4844
fax: 201-575-4011

GIBCO BRL
P.O. Box 68
Grand Island, NY 14072-0068
phone: 800-828-6686
fax: 800-331-2286

Harlan Biotechnologies
P.O. Box 29176
Indianapolis, IN 46229
phone: 619-450-3112
fax: 317-894-1840

Jackson Immunoresearch Laboratories
872 West Baltimore Pike
P.O. Box 9
West Grove, PA 19390
phone: 800-367-5296
fax: 215-869-0171

Marsh
7171 Patterson
Garden Grove, CA 92641
phone: 800-283-2326

Molecular Probes
4849 Pitchford Avenue
Eugene, OR 97402-9144
phone: 800-438-2209
fax: 800-438-0228

Narishige USA, Inc.
One Plaza Road
Greenvale, NY 11548
phone: 516-621-4588, 800-445-7914
fax: 516-621-5081

National Diagnostics
1013-1017 Kennedy Blvd.
Manville, NJ 08835
phone: 800-526-3867

Panasonic Industrial Co.
One Panasonic Way
Seacaucus, NJ 07094
phone: 201-348-7620

Pelikan
(try your local stationery store first)
Pelikan AG
D3000 Hannover 1
Germany

Polysciences, Inc.
400 Valley Road
Warrington, PA 18976
phone: 800-523-2575
fax: 800-343-3291

R&D Products
1808 Harmon Street
Berkeley, CA 94703
phone: 510-547-6464

S & M Microscopes, Inc.
1064 W. Parkhill Drive
Boise, ID 83702
phone: 208-338-1243

Sigma Chemical Co. and
Sigma Immunochemical Co.
P.O. Box 14508
St. Louis, MO 63178
phone: 800-325-3010
fax: 800-325-5052

Stoelting Co.
620 Wheat Lane
Wood Dale, IL 60191
phone: 708-860-9700
fax: 708-860-9775

Thomas Scientific
99 High Hill Road at I-295
P.O. Box 99
Swedesboro, NJ 08085-0099
phone: 800-345-2100
fax: 609-467-3087

USA/Scientific Plastics
P.O. Box 3565
Ocala, FL 34478
phone: 800-LAB-TIPS (800-522-8477)
fax: 904-351-2057

Vector Laboratories
30 Ingold Road
Burlingame CA 94010
phone: 800-227-6666
fax: 415-697-0339

Vincent Associates
1255 University Avenue
Rochester, NY 14607
phone: 800-828-6972

VWR Scientific
P.O. Box 39396
Denver, CO 80239
phone: 801-972-6400

Wild Heerbrugg
9435 Heerbrugg
Switzerland
phone: (071) 70 31 31
fax: (071) 70 31 70

World Precision Instruments, Inc.
175 Sarasota Center Blvd.
Sarasota, FL 34240
phone: 813-371-1003
fax: 813-377-5428

Index

Index

A

B

C

D

E

F

G

H

I

S

T

V